Food and Drug Regulation in an Era of Globalized Markets

Food and Drug Regulation in an Era of Globalized Markets

Edited by

Sam F. Halabi
The O'Neill Institute for National and Global Health Law, Georgetown University, Washington, DC and The University of Tulsa College of Law, Tulsa, OK, USA.

ELSEVIER

AMSTERDAM • BOSTON • HEIDELBERG • LONDON
NEW YORK • OXFORD • PARIS • SAN DIEGO
SAN FRANCISCO • SINGAPORE • SYDNEY • TOKYO
Academic Press is an imprint of Elsevier

Academic Press is an imprint of Elsevier
125 London Wall, London, EC2Y 5AS, UK
525 B Street, Suite 1800, San Diego, CA 92101–4495, USA
225 Wyman Street, Waltham, MA 02451, USA
The Boulevard, Langford Lane, Kidlington, Oxford OX5 1GB, UK

Library of Congress Cataloging-in-Publication Data
A catalog record for this book is available from the Library of Congress

British Library Cataloguing in Publication Data
A catalogue record for this book is available from the British Library

ISBN: 978-0-12-802311-2

Publisher: Nikki Levy
Senior Acquisitions Editor: Nancy Maragioglio
Editorial Project Manager: Billie Jean Fernandez
Production Project Manager: Chris Wortley
Designer: Victoria Pearson

Typeset by SPi Global, India

Contents

Contributors

Elle Alexander
Policy Analyst, The Vitality Institute

Ms. Alexander's work at the Vitality Institute centers on research, and engaging with stakeholders in the business, NGO, and scientific community. She leads the Vitality Institute's Research and Development partnerships and projects to bring the best evidence on chronic disease risk reduction into practice. Ms. Alexander completed her MSPH in human nutrition at Johns Hopkins Bloomberg School of Public Health. While at John Hopkins, she worked on the Baltimore Healthy Eating Zones (BHEZ) project to promote healthy eating behaviors in youth through interventions in corner stores and recreation centers. Following her MSPH, Ms. Alexander worked at PepsiCo as a policy analyst on the global health and agriculture policy team to promote collaborative solutions for improving nutrition and reducing chronic disease risk factors. Ms. Alexander holds a Bachelor of Science in psychology, and an International Relations certificate from the University of Massachusetts Amherst.

Shira Epstein
Associate, Latham & Watkins, LLP

Shira Epstien is an associate in the Washington, D.C. office of Latham & Watkins. Ms. Epstien specializes in litigation, investigations, and compliance matters in the healthcare industry. Ms. Epstien's practice focuses on counseling hospital, pharmaceutical, medical device, food, and other related industry clients with respect to regulatory issues involving the Centers for Medicare and Medicaid Services, and the Food and Drug Administration. She has also advised clients in matters pertaining to violations of the False Claims Act.

Ms. Epstien received her J.D. from Georgetown University Law Center, and her Master of public policy degree from the Harvard Kennedy School. In addition to her legal training at Georgetown, Ms. Epstien studied transnational law at the Center for Transnational Legal Studies in London.

Aliza Glasner
Institute Associate, O'Neill Institute for National and Global Health Law

Aliza Glasner is an O'Neill Institute Associate. Ms. Glasner's primary focus is on food and drug law. Before joining O'Neill, she was the manager of Product Development at the Food and Drug Law Institute in Washington, D.C. There, she worked in collaboration with industry and government experts to develop programs and publications addressing legal, policy, and regulatory issues in the food and drug sector. Ms. Glasner was also responsible for planning international events, including FDLI's inaugural conference in Beijing, China, which focuses on issues related to trade, compliance, and safety in the food and drug sector. She previously practiced law at Cadwalader, Wickersham & Taft in New York. Ms. Glasner holds a law degree from Fordham University and a B.A. in international affairs from the George Washington University.

Jesse Goodman
Director, Center on Medical Product Access, Safety and Stewardship (COMPASS), Georgetown University Medical Center; Former Chief Scientist, US Food and Drug Administration

Jesse L. Goodman, MD, MPH, is the founding director of the Center on Medical Product Access, Safety and Stewardship (COMPASS) at Georgetown University Medical Center. The Center studies three broad areas of U.S. and global public health importance: medical product safety and security across the global manufacturing supply chain, antimicrobial drug resistance, and access to lifesaving medical products. Goodman, a member of the Institute of Medicine, was most recently the chief scientist for the U.S. Food and Drug Administration (FDA), and is a renowned expert in infectious diseases. In addition to leading COMPASS, Dr. Goodman is an attending physician at the Washington D.C. VA Medical Center and MedStar Georgetown University Hospital. Prior to joining the FDA, Dr. Goodman was director of the division of infectious diseases and professor of medicine at the University of Minnesota, where his National Institutes of Health-funded laboratory isolated and characterized the causative agent of a new tick-borne disease, known as human granulocytic anaplasmosis. He received his AB in biology from Harvard College, his MD from the Albert Einstein College of Medicine, and his MPH from the University of Minnesota.

Lawrence O. Gostin
Linda D. & Timothy J. O'Neill Professor of Global Health Law; Faculty Director, O'Neill Institute for National & Global Health Law; Director, World Health Organization Collaborating Center on Public Health Law & Human Rights, University Professor

Lawrence O. Gostin is professor of medicine at Georgetown University, professor of public health at the Johns Hopkins University, and director of the Center for Law & the Public's Health at Johns Hopkins and Georgetown Universities. Prof. Gostin holds a number of international academic professorial appointments: visiting professor (faculty of medical sciences) and research fellow (Centre for Socio-Legal Studies) at the University of Oxford, United Kingdom; the Claude Leon Foundation Distinguished Scholar and visiting professor at the University of Witwatersrand, Johannesburg, South Africa; and the Miegunyah distinguished visiting fellow, and founding fellow of the Centre for Advanced

Studies (Trinity College), University of Melbourne. Prof. Gostin serves as secretary and a member of the Governing Board of Directors of the Consortium of Universities for Global Health. Prof. Gostin is the director of the World Health Organization Collaborating Center on Public Health Law & Human Rights. He also serves on the WHO Director-General's Advisory Committee on Reforming the World Health Organization. In 2007, the WHO Director-General appointed Prof. Gostin to the International Health Regulations (IHR) Roster of Experts and the Expert Advisory Panel on Mental Health.

Prof. Gostin holds numerous editorial appointments in leading academic journals throughout the world. His principal position is the health, law, and ethics editor, contributing writer, and columnist for the *Journal of the American Medical Association*. He is also founding editor-in-chief of *Laws* (an international, open-access, law journal). He was formerly the editor-in-chief of the *Journal of Law, Medicine & Ethics*. He holds a B.A. from the State University of New York, Brockport; a JD from Duke University; an LLD (Hon.) from the State University of New York; and an LLD (Hon.) from the University of Sydney. Prof. Gostin, an elected lifetime member of the Institute of Medicine/National Academy of Sciences, serves on the Board on Health Sciences Policy, the Human Subjects Review Board, and the Committee on Science, Technology, and Law. He currently chairs the IOM Committee on Global Solutions to the Challenge of Counterfeit Medicines.

J. Benneville (Ben) Haas
Partner, Latham & Watkins

Ben Haas practices in Latham & Watkins' Washington, D.C. office, where he is a corporate partner in the healthcare and life sciences practice group. Mr. Haas focuses his practice on regulatory, transactional, and legislative matters involving the medical device, pharmaceutical, biotechnology, tobacco, food, cosmetic, and dietary supplement industries. Mr. Haas has counseled clients on a wide variety of regulatory matters involving the U.S. Food and Drug Administration (FDA). In the transactional arena, Mr. Haas provides counsel on corporate compliance, mergers and acquisitions, and other corporate transactions. He regularly assists pharmaceutical and medical device companies in initiating, managing, and completing large, international clinical trials. He has counseled clients regarding clinical trial strategy; contracting with and managing clinical research organizations, other clinical trial vendors, and clinical trial sites and investigators in the United States, and in dozens of overseas jurisdictions; and communicating with U.S. and foreign regulatory authorities, as well as institutional review boards and ethics committees, both domestic and foreign. Mr. Haas has also drafted and negotiated several types of agreements for healthcare clients, including manufacturing and supply, licensing, joint development, consulting and marketing, and distribution contracts. Mr. Haas has conducted regulatory diligence for major public and private financings, mergers, and other transactions in the device, pharmaceutical, and dietary supplement industries. Mr. Haas also counsels clients on legislative and administrative rulemaking proceedings relating to the FDA, including issues resulting from the implementation of the Family Smoking Prevention and Tobacco Control Act of 2009, the Food and Drug Administration Amendments Act of 2007, the Medical Device User Fee and Modernization Act of 2002, the Bioterrorism Act of 2002, and the Medicare Prescription Drug Improvement and Modernization Act of 2003. He has worked extensively with trade associations and other clients to address specific regulatory issues in matters before several federal agencies, including the Environmental Protection Agency, Federal Trade Commission, Department of Transportation, U.S. Department of Agriculture and Drug Enforcement Administration. Mr. Haas has authored, and co-authored, articles on FDA regulatory matters, including articles in the Food and Drug Law Institute's *Update Magazine* and the *Pharmaceutical Law & Industry Report*, and has been a featured speaker and lecturer on FDA regulatory and policy matters at industry-sponsored events.

Sam F. Halabi
Associate Professor, The University of Tulsa College of Law and Scholar, O'Neill Institute for National and Global Health Law

Professor Halabi is a scholar of national and global health law with specializations in food, health services, and pharmaceutical business organizations. Between 2008 and 2010, he served as fellow of the O'Neill Institute for National and Global Health Law at Georgetown University. In 2008, he advised the Presidential Health Care Policy Working Group on current and proposed approaches to international food and drug inspections. In 2010, Professor Halabi addressed a special committee of the United Nations on the relationship between decentralized health care systems and social inclusion. He serves as an adviser to the National Foundation for the Centers for Disease Control and Prevention, and the World Health Organization. He also serves as vice-chair of the *Food and Drug Law Journal* editorial board, as well as a referee for the journals *Global Public Health*, *Health and Human Rights*, and the *Journal of Law, Medicine & Ethics*. Before earning his JD from Harvard Law School, Professor Halabi was awarded a British Marshall scholarship to study in the United Kingdom where he earned an M.Phil in international relations from the University of Oxford (St. Antony's College). During the 2003-04 academic year, he served as a Rotary International Ambassadorial Scholar at the American University of Beirut.

Margaret Hamburg
Commissioner, U.S. Food and Drug Administration (FDA)

Margaret A. Hamburg became the 21st commissioner of food and drugs on May 18, 2009. The second woman to be nominated for this position, she is an experienced medical doctor, scientist, and public health executive. Dr. Hamburg graduated from Harvard Medical School and completed her residency in internal medicine at what is now New York-Presbyterian Hospital/Weill Cornell Medical Center. She conducted research on neuroscience at Rockefeller University, studied neuropharmacology at the National Institute of Mental Health, and later focused on AIDS research as assistant director of the National Institute of Allergy and Infectious Diseases. From 2005 to 2009, Dr. Hamburg was the senior scientist at the Nuclear Threat Initiative, a foundation dedicated to reducing the threat to public safety from nuclear, chemical, and biological weapons. From 2001 to 2005, as the foundation's vice president for biological programs, she advocated for broad reforms to confront the dangers of modern bioterrorism, as well as the threats of naturally occurring infectious diseases, such as pandemic flu. In 1997, Dr. Hamburg accepted the position of assistant secretary for policy and evaluation in the U.S. Department of Health and Human Services. In 1994, she was elected to the membership in the Institute of Medicine, one of the youngest persons to be so honored. From 1991 to 1997, Dr. Hamburg served as commissioner of the New York City Department of Health and Mental Hygiene. In this position, she improved services for women and children, promoted needle-exchange programs to reduce the spread of HIV, and initiated the nation's first public health bioterrorism defense program. Her most celebrated achievement was curbing the spread of tuberculosis, which resurged as a major public health threat in the 1990s. As a result of Dr. Hamburg's reforms, New York City's TB rate fell significantly over a 5-year span. Her innovative approach, which included sending health care workers to patients' homes to make sure they completed the drug regimen, is now a model for health departments worldwide.

Martin Harvey-Allchurch
Head of Communications Service, European Medicines Agency

Martin Harvey Allchurch is a lawyer by training, with degrees in English, Scottish, European and international law. He joined a public affairs lobbyist in Brussels in 1992, specializing in regulated industries and the sports sector. Mr. Harvey-Allchurch started at the European Medicines Agency in 1995, initially as part of the legal team. He headed up the Office of the Executive Director from 2004 to 2012, with a range of policy and management responsibilities, and led the creation of the agency's new communications service from 2012 to 2014. He joined the international affairs staff in November 2014.

Sabine Haubenreisser
European Medicines Agency, Liaison Official to US FDA

Sabine Haubenreisser holds a master's in pharmaceutical science from the Johann Wolfgang Goethe-Universität in Frankfurt am Main; a PhD in pharmacology from the Pharmakologisches Institut für Naturwissenschaftler in Frankfurt am Main; and a certificate in epidemiology from the London School of Hygiene and Tropical Medicine. In 1997, she joined the European Medicines Agency (EMA), where she held positions as support to the Head of Human Unit, as project manager in the Sector of New Chemical Substances, as secretary to the Mutual Recognition Facilitation Group, and, from 2001 to 2009, as specialized group leader of the Team Anti-Infectives, Immunology, Gastroenterology and Musculoskeletal in the Post-Authorisation Human Unit.

From 2009 to 2012 she worked as officer for External Communications and Media in the office of the executive director, where she developed communication strategies and prepared EMA public communications. In May 2012, she was appointed as EMA liaison official at the U.S. FDA, and is based in White Oak, Silver Spring, MD. In this role, she facilitates cooperation between the European Regulatory Network for Medicines and the Food and Drug Administration, supporting the regulatory and scientific understanding and convergence between both regions.

Lisa Heinzerling
William J. Brennan, Jr. Professor of Law, Georgetown University Law Center
Former Associate Administrator, US Environmental Protection Agency Office of Policy, 2009-10

Lisa Heinzerling is professor of law at Georgetown University. Her specialties include environmental and natural resources law, administrative law, the economics of regulation, and food and drug law. From January 2009 to July 2009, Heinzerling served as senior climate policy counsel to the administrator of the Environmental Protection Agency. Then, from July 2009 to December 2010, she served as associate administrator of the EPA's Office of Policy. In 2008, she served as a member of President Obama's EPA transition team. Professor Heinzerling has been a visiting professor at Harvard Law School, Vermont Law School, and Yale Law School. She lectures frequently on environmental law and other topics, both in the U.S. and around the world. She has published several books, including a leading casebook (with Zygmunt Plater, and others) on environmental law; a casebook (with Mark Tushnet) aimed at introducing first-year law students to the regulatory and administrative state; and a widely cited critique of the use of cost-benefit analysis in environmental policy (*Priceless: On Knowing the Price of Everything and the Value of Nothing*, co-authored with Frank Ackerman). The Yale Environmental Law Association and Yale Center for Environmental Law and Policy recently gave her their inaugural award for innovative and inspiring scholarship in environmental law.

Ching-Fu Lin
Institute of Law for Science and Technology, National Tsing Hua University, Hsinchu, Taiwan

Ching-Fu Lin holds a double degree in law (LLB) and chemical engineering (BS) from the National Taiwan University. In 2010, he earned his LLM from Harvard Law School, where he is currently a candidate for the SJD. He was Peter Barton Hutt Student Fellow at the Petrie-Flom Center for Health Law Policy, Biotechnology, and Bioethics, Harvard Law School in 2013–2014. Lin previously served as visiting fellow at the Graduate Institute of International and Development Studies (IHEID) in Geneva, as well as researcher and associate journal editor at the Asian Center for WTO & International Health Law and Policy; he recently joined the Institute of Law for Science and Technology, National Tsing

Hua University, Taiwan. His areas of research include food safety law and policy, international trade law, global health law, and international relations theory. His legal scholarship has appeared in numerous journals and edited collections, among them are: "Global Food Safety: Exploring Key Elements for an International Regulatory Strategy," in the *Virginia Journal of International Law*, 2011; "SPS-Plus and Bilateral Treaty Network: A 'Global' Solution to the Global Food Safety Problem?" in the *Wisconsin International Law Journal*, 2012; "Reassessing the Limits of the Codex Alimentarius Commission," in the *Columbia Science and Technology Law Review*, 2013; and "Public-Private Interactions in Global Food Safety Governance," in the *Food and Drug Law Journal*, 2014.

John Monahan
Senior Advisor to the President for Global Health and Senior Fellow, McCourt School of Public Policy, Georgetown University; Former Counselor to the Secretary and Director, Office of Global Health Affairs, US Department of Health and Human Services (HHS)

John T. Monahan is the senior advisor for Global Health to Georgetown University President John J. DeGioia, and a senior fellow at the University's McCourt School of Public Policy. Most recently, Mr. Monahan served as special advisor for Global Health Partnerships at the United States Department of State, where he led the Obama Administration's successful diplomatic campaign to reform the Global Fund to Fight AIDS, Tuberculosis, and Malaria. Mr. Monahan has also served as: counselor to Secretary Kathleen Sebelius at the Department of Health and Human Services (HHS), director of HHS's Office of Global Health Affairs, and as a member of the Obama-Biden Transition Team. In the Clinton Administration, Mr. Monahan served at HHS as principal deputy assistant secretary for Children and Families, supporting the implementation of federal welfare reform legislation; and as director of Intergovernmental Affairs, managing scores of health and welfare reform waiver negotiations with governors' offices and senior state officials. He also worked for more than six years as a senior fellow with the Annie E. Casey Foundation, focusing on innovative investment strategies for low-income families and communities. He was the founding executive of Georgetown's O'Neill Institute for National and Global Health Law. He is a *cum laude* graduate of Georgetown College and Georgetown Law.

The Pew Charitable Trusts

The Pew Charitable Trusts is an independent nonprofit global research and public policy organization, operated as a non-partisan, non-governmental organization dedicated to serving the public. Pew's mission is to improve public policy by conducting rigorous analysis, linking diverse interests to pursue common cause and insisting on tangible results; inform the public by providing useful data that illuminate the issues and trends shaping our world; and to invigorate civic life by encouraging democratic participation and strong communities. From 2010 to 2013, The Pew Charitable Trusts conducted a comprehensive assessment of the federal food additives regulatory program. Relying on a process that engaged stakeholders, Pew examined food additive issues in partnership with the food industry, the public interest community, and the federal government, including the U.S. Food and Drug Administration. Pew held five expert workshops and published six reports in peer-reviewed journals. The chapter included herein summarizes Pew's findings and provides recommendations to address the problems identified.

Gordon Schatz
Executive Vice President, Healthcare, China Trade Strategies, Washington, DC, USA

Gordon Schatz has served as Reed Smith's China life science integration partner, since 2009, working with medical device and drug companies through Reed Smith's offices in Beijing and Shanghai. Mr. Schatz represents health care manufacturers and providers of innovative products and services on regulatory, reimbursement, and compliance issues. He co-edited FDLI's *Compilation of Chinese Food and Drug Administration Regulations on Drugs and Devices*, with Professor Wang Chenguang, from the Tsinghua School of Law, Beijing. In the United States, Gordon is a recognized expert on Medicare coverage, coding, and payment for specialized medical services, including cardiology, radiology, nuclear medicine, and minimally invasive surgery. He is a graduate of Temple Law School and Haverford College.

Marc Scheineson
Partner, Chair, Food and Drug Law Practice Group, Alston & Bird, LLP

Marc J. Scheineson is a partner in the Washington, D.C. office of Alston & Bird. He heads the firm's food and drug law practice. Mr. Scheineson previously served as associate commissioner for Legislative Affairs at the FDA. He came to Washington, D.C. to work as legislative assistant and counsel to Rep. Bill Gradison (R-OH), the ranking member of the Health Subcommittee of the House Ways and Means Committee. In addition, he worked as a regulatory FDA lawyer for another national law firm, and senior vice president of Ketchum Communications, and as a principal in its government relations unit. Mr. Scheineson is a former co-chairman of the American Bar Association's Committee on Food and Drug Law. He chaired the Task Force on FDA Reform, and is a former chair of the Young Lawyers Section of the Bar Association of the District of Columbia. He received his BA and JD from the University of Cincinnati and its College of Law, and his LLM from the Georgetown University Law Center.

Adriana Selwyn
Policy Analyst, The Vitality Institute

Ms. Selwyn's research and development work at the Vitality Institute focuses on building evidence-based health promotion and chronic disease prevention programs among working-age Americans. She obtained her Bachelor of Science (Hons) in nutrition from the University of Sydney, Australia. She worked as a dietitian in a Sydney-based tertiary hospital providing medical nutrition therapy to patients in a range of areas, including: oncology, gerontology, cardiology, and renal medicine. Ms. Selwyn also worked with patients in the community, treating a range of non-communicable diseases. Following this, Ms. Selwyn moved into public health research, where she focused on a population-wide salt reduction program in her work with the Food Policy group at The George Institute for Global Health.

Vince H. Suneja
Chief Executive Officer, TwoFour Insight Group, LLC; Former First Secretary, Market Access & Compliance Attaché at the U.S. Embassy in New Delhi, India

Vince H. Suneja is CEO and co-founder of TwoFour Insight Group, LLC, responsible for the financial management, operational oversight, and strategic direction of the life sciences company's operations. Before starting TwoFour Insight Group, Mr. Suneja served as the first secretary, Market Access & Compliance attaché at the U.S. Embassy in New Delhi, India, where he opened up the Embassy's first Market Access & Compliance Office on behalf of the U.S. Department of Commerce's International Trade Administration (ITA). Mr. Suneja earned his master's in business administration from the University of Pittsburgh's Katz School of Business; his Master of Laws in global health law; a World Trade Organization certificate from Georgetown University Law's Institute of International Economic Law Center, and a Juris Doctor from Ohio Northern University, Petit College of Law.

Chris Waldrop
Director, Food Policy Institute, Consumer Federation of America

Christopher A. Waldrop is the director of the Food Policy Institute at the Consumer Federation of America; an association of nearly 280 non-profit consumer organizations that seeks to advance the consumer interest through research, advocacy, and education. Mr. Waldrop oversees the research, analysis, advocacy, and media outreach for all food policy activities at the Institute. He regularly monitors food safety activities at USDA, FDA, and in Congress, where he advocates for strong food safety protections for consumers. Mr. Waldrop also coordinates the Safe Food Coalition, a group of consumer, trade union, and foodborne illness victim organizations dedicated to reducing foodborne illness by improving government food inspection programs.

Mr. Waldrop serves on the USDA's National Advisory Committee on Meat and Poultry Inspection, which advises the secretary of Agriculture on matters affecting federal and state inspection program activities. From 2010 to 2013, he served on the FDA's Food Advisory Committee, which advises the commissioner on emerging food safety, food science, nutrition and other policy-related health issues. In 2009 and 2011, Mr. Waldrop served on two National Academy of Sciences panels charged with reviewing USDA food inspection programs.

Mr. Waldrop holds an advertising degree from Texas Tech University, and a Master of Public Health degree from Johns Hopkins University.

Derek Yach
Senior Vice President of the Vitality Group; Former Executive Director for Noncommunicable Diseases and Mental Health, World Health Organization

Derek Yach is senior vice president (SVP) of the Vitality Group, part of Discovery Holdings Ltd, where he leads the Vitality Institute. Prior to that he was SVP global health and agriculture policy at PepsiCo, where he supported portfolio transformation and led engagement with major international groups, and new African initiatives at the nexus of agriculture and nutrition. He has headed global health at the Rockefeller Foundation, has been a professor of global health at Yale University, and is a former executive director for Noncommunicable Diseases and Mental Health of the World Health Organization (WHO). At the WHO, he served as cabinet director under Director-General Gro Harlem Brundtland, where he led the development of WHO's Framework Convention on Tobacco Control and the Global Strategy on Diet and Physical Activity. Dr. Yach established the Centre for Epidemiological Research at the South African Medical Research Council. He has authored, or co-authored, over 200 articles covering the breadth of global health. Dr. Yach serves on several advisory boards, including those of the Clinton Global Initiative, the World Economic Forum, the NIH's Fogarty International Centre, and PepsiCo's scientific advisory board. His degrees include an MBChB from the University of Cape Town, a BSc (Hons Epi) from the University of Stellenbosch, and an MPH from the Johns Hopkins Bloomberg School of Public Health.

Foreword

The U.S. Food and Drug Administration (FDA) has been protecting consumer health and safety for more than 100 years, but the forces of globalization have required a significant reexamination of the FDA's roles and responsibilities. There are now few products regulated by the FDA that are exclusively domestic in origin, manufacture, and distribution. Products enter our country through complex supply chains, involving more than 300,000 facilities around the world and some 150 countries. We must take a global view.

There is no aspect of consumer products untouched by these movements: manufacturing, processing, packaging, and distributing are all part of geographically diverse food and drug markets. As a result, we need a new level of engagement with our appropriate counterpart regulatory authorities as well as the industries that the FDA regulates. In addition, we need new programs and policies that reflect this engagement.

To some extent, this means harnessing novel technologies and methods of analysis to strengthen existing approaches. For example, regulators still screen products at points of entry, but now we use risk-based analytics and sophisticated algorithms to categorize the shipments coming into ports of entry (land, sea, and airport), identify the nature of the product, and apply aggregate knowledge about risks from the general to the specific. For example, the past history of a company's quality performance or product safety information from a fellow regulator can be integrated into our routine risk screening. Similarly, at a given point in time, certain environmental events may have affected the safety or integrity of a given product and we are in a better position to assess that possibility. The result is a more efficient system that allows the FDA to readily classify shipments to determine whether products require additional inspection: "remove the container from the ship," "stop the truck and inspect," or, of course, "let them through."

Improvements in the domestic inspection sphere are important, but regulators must also move closer to where the products and their components are actually produced and the pathways they follow to our land. This requires a presence in many countries and regions around the world and new levels of collaboration with industry and regulators. Over the past decade, the FDA has established multiple foreign offices around the world. These offices can serve as a hub for inspections of foreign facilities and manufacturing sites, but they also represent an important locus to enhance in-country/regional engagement with key partners. In addition, we have been involved in a set of bilateral and multilateral agreements with other countries to share information and cooperate on specific activities, as well as working on international cooperation and harmonization through a number of international organizations. Clearly, such arrangements have played an important role, but the truth is that this is a patchwork approach; additionally, pursuing a series of "one-off" bilateral and multilateral agreements requires substantial time and resources to negotiate, implement, and regularly review.

Our modern, highly interconnected world demands an approach that engages all relevant stakeholders in a systematic and continuing manner. As many countries struggle to address the challenges of globalization and increasingly complex supply chains, there is a growing commitment to put in place a new system of global governance. This vision of a new governance mechanism has generated huge interest and enthusiasm from regulators, regulated industry, and relevant stakeholders who want to work together to put in place a new global system that will more fully ensure consumer safety.

Finding the right model or mix of models for this system is a challenge, but the FDA has actively explored possibilities derived from existing global government mechanisms in the financial sector, the airline industry, and trade regulation, among many others. Whatever its form, we need an institutionalized, umbrella mechanism that serves as a nexus between partners. It is frequently said that industry resists regulation. But firms often prefer the predictability that strong regulatory oversight provides and they have played an important part in many of the efforts now under way. The food industry, for example, has worked with the World Bank and other partners to try to help strengthen the regulatory capacity in developing countries to ensure food safety.

Of course, any global governance mechanism must be flexible enough to accommodate national differences. Countries have very different sovereign laws and practices that define important aspects of national legal-regulatory frameworks. Laws regulating confidential commercial information and information processes, for example, vary significantly, as do the specifics of regulatory pathways and requirements. In addition, there are significant differences in history, culture, and resources that have shaped national regulatory capacity and approaches. Ultimately, this means that not every national regulatory authority can or should look like the FDA, but our goal is to find ways to converge at public health and scientific outcomes and build sufficient confidence in those outcomes that information may be effectively shared and used.

Building a global governance approach makes sense for multiple reasons, including efficiency and burden sharing. If hundreds of thousands of facilities all over the world are contributing products that reach similarly diverse and far-reaching markets and consumers, information sharing allows the FDA and fellow regulatory authorities to develop a more coordinated approach that can raise standards, ensure quality, and better leverage expertise and fiscal resources. Such an approach can also reduce the burden on both regulators and industry by alleviating the need for multiple different regulatory authorities to perform similar functions like inspections over and over again. Information sharing can clarify the optimal role for each authority in the global context. It also gives the FDA the opportunity to work in new ways and establish vital new relationships.

A corollary to the aforementioned governance effort is investment in global regulatory capacity. For both humanitarian and health reasons, it is in our own vital interest to help build regulatory capacity in developing countries. In fact, if products that are coming into this country are made in countries with more robust regulatory oversight, everyone benefits. In our view, investments to strengthen regulatory capacity in developing countries are a win-win-win: improving consumer health safety in the United States and in other countries, while also offering meaningful health and economic development benefits to the countries concerned. That is why the FDA has been working with partners like the World Bank, multilateral organizations, development agencies, and with philanthropies to develop programs that enhance regulatory capacity in other countries.

While not traditionally viewed as part of development investment portfolios, building regulatory capacity should be included as a development goal because strong regulatory capacity improves domestic health and productivity and supports reliable industry and economic growth. For example, such efforts can increase access to safe, effective medical products manufactured and distributed with quality standards, improving health and health care systems. Similarly, in many developing countries, where food supplies are limited, strong regulatory capacity ensures that food supplies are not lost to foodborne outbreaks or other contamination, and people are spared preventable disease. For both food and medical products, stronger regulatory capacity in turn enhances domestic industry and strengthens growth and trade potential. It is clearly in the U.S.'s interest to build regulatory capacity because so many products destined for the U.S. market originate in or pass through developing countries and we can better ensure safety and quality of standards with strong partners.

Finally, investment in strong regulatory capacity is an investment in key aspects of national and international security. Today, we are all vulnerable to unintended or deliberate adulteration of vital products if we do not recognize and respond to the realities of our globalized world. At the FDA we are working hard to be part of a new era of global safety and regulatory oversight that will promote and protect the health of the public in our country and beyond.

Margaret A. Hamburg
Commissioner, U.S. Food and Drug Administration

Introduction

Sam F. Halabi[*,†]
*The University of Tulsa College of Law, †O'Neill Institute for National and Global Health Law, Washington, District of Columbia, USA

Globalization—in essence the ability to sell the same or similar products and services worldwide and the international agreements promoting the same—has served as the rubric for sweeping analyses as to how economic and social structures on all continents are undergoing transformational change to subtle and profound influences on art and culture. Although the term dates to 1983, its deeper relevance has only been developed over the last decade or so. Between 2002 and 2004, Nobel Laureate Joseph Stiglitz and *Financial Times* Chief Economic Commentator Martin Wolf published their dueling theses criticizing and praising globalization's effect on societies in both the industrialized and developing worlds. Tom Friedman wrote *The World is Flat*, exploring vignettes that illustrated the broad trends Stiglitz and Wolf outlined. The *New Yorker*'s Patrick Keefe elaborated globalization's trends both top-down (in *Chatter*, how governments increasingly rely upon each other to eavesdrop on their citizens) and bottom-up (in *The Snakehead*, how globalized labor markets facilitated modern human trafficking), while Franklin Foer explained globalization's effect on players, teams, and spectators of the world's most popular sport in *How Soccer Explains the World.*

The aforementioned accounts are, for the most part, descriptive, identifying the movements afoot generating both opportunities to raise the global standard of human quality of life as well as deep, vexing challenges in securing what gains have been achieved from criminal and illicit networks that now operate more widely than ever before, in addition to deficiencies in oversight as capital, labor, and products surge through ever more-open borders.[1] The challenge facing consumers, firms, regulators, and scholars alike is how to coordinate and reconstruct existing regulatory structures to facilitate and promote the positive aspects of globalization while containing and minimizing the costs associated with criminal activity and regulatory gaps. This challenge has prompted important empirical and theoretical work, even if the challenge remains as pressing today as ever. Walter Mattli's and Ngaire Woods's *The Politics of Global Regulation* gathered and classified perspectives on efforts to regulate shipping, sovereign debt, corporations, labor rights, human rights, and the environment through "a plethora of non-state and public-private governance arrangements."[2] Daniel Drezner analyzed global governance regimes

1. Wolf, M., 2004. Why Globalization Works. Yale University Press; Stiglitz, J., 2003. Globalization and Its Discontents. W. W. Norton and Company.
2. Mattli, W., Woods, N., 2009. The Politics of Global Regulation, 44. Princeton University Press.

for Internet classifications, financial institutions, genetically modified organisms, and intellectual property rights in support of his argument that, whatever globalization's effect to date, global regulatory structures represent broadly similar analogs to domestic counterparts, including the disproportionate importance of regulatory regimes prevailing in larger economies like the United States and the European Union.[3]

Because of the early and prominent role pharmaceutical and food firms played in establishing global commercial footprints, supporting trade and investment liberalization initiatives, and influencing individual and public health on a large scale, global regulation over food and medicines has always held a central position in debates about globalization, the nature and structure of supranational regulation, and international regulatory cooperation aimed at managing globalization's benefits and costs. The legal confrontation between the South African government and the world's major pharmaceutical firms represented a watershed moment in the global regulatory balance between intellectual property protections guaranteed in the 1995 Agreement on the Trade-Related Aspects of International Trade (TRIPS) and ensuring access to medicines as part of states' constitutional obligations. The Doha Declaration on the TRIPS Agreement and Public Health effectively amended the global regulatory regime for intellectual property rights in the public health emergency context. The World Trade Organization's Agreement on Sanitary and Phytosanitary Measures gave decisive regulatory standard-setting authority to the joint WHO-FAO Codex Alimentarius Commission. The realization that common regulatory standards were crucial to global access to medicines drove some of the earliest regulatory harmonization efforts such as the International Conference on Harmonization (of Technical Requirements for Registration of Pharmaceuticals for Human Use).

Academic analysis of global regulatory regimes and their restructuring has continued apace, although with the conventional tendency to specialize. Since 2005, the U.S. Institute of Medicine has, through its Forum on Drug Discovery, Development, and Translation, held a number of workshops and published several studies related to the problems globalization has caused for the integrity and oversight of the pharmaceutical supply chain. In 2009, professors Jennifer Clapp and Doris Fuchs edited a volume committed to the evolution of "corporate power in agrifood governance" exposing some of the regulatory weaknesses at international food safety organizations like the Codex Alimentarius Commission. For the most part, however, broad scholarly treatments of globalized food and drug regulation have attempted to draw conclusions based on narrow, sometimes comparative case studies like Arthur Daemmrich's *Pharmacopolitics* or Peter Oosterveer's and David Sonnenfeld's *Food, Globalization and Sustainability*.

3. Drezner, D. W., 2007. All Politics is Global: Explaining International Regulatory Regimes. Princeton University Press.

Globalized regulation of food and drugs is similarly analyzed within volumes that approach regulation generally and include food and drugs as having experiences in common with financial, telecommunications, or transport sectors like John Braithwaite's and Peter Drahos's *Global Business Regulation* or Maki Ememura's and Rika Fujioka's *Comparative Responses to Globalization*. There are, of course, innumerable articles, essays, and monographs as well as firm-, regulator- and think tank-generated reports discussing the globalization of food and drug regulation.

This book contributes to the existing discussion by bringing together in a single volume not only firms, regulators, and academics (like the IOM), scholars from different disciplines (like Mattli and Woods), or scholars from different geographic regions (like Clapp and Fuchs), but also the full range of perspectives from stakeholders in the global supply chains for food and medicines: consumer associations, attorneys for pharmaceutical firms bringing drugs to market through regulated activities taking place on multiple continents, national and international regulators, firm-side participants in public-private partnership regulatory initiatives, scholars of medicine, regulatory science, political science, international relations, and public health, as well as researchers based at both universities and major nongovernmental organizations. Through this collaborative dialog, this volume aims to strike a heretofore missing balance between university-based researchers (who are almost always disproportionately represented in the secondary literature), regulators, firms (whose work on regulation is, while influential, often proprietary and nonpublic), and consumers (who are often depicted or theorized by the three aforementioned stakeholders rather than representing themselves). Together, these perspectives will convey not only what is under way in terms of the changing nature and structure of food and drug regulation as markets globalize but also identify the most important regulatory gaps, public health risks, and promising solutions that balance the interests of private for-profit firms, resource constrained regulators, and consumers.

Following a discussion of the relationship between globalization and the challenges facing regulation of food and pharmaceuticals more broadly, this introductory chapter presents a conceptual framework to explain the mode of interdisciplinarity it adopts. The topics covered by the contributors to this volume have been selected to allow as comprehensive an understanding as possible of the most significant regulatory challenges facing the widest range of stakeholders. They include approaches to falsified and substandard medicines introduced into global pharmaceutical supply chains; management of multiregional clinical trials; changes in the national regulatory approaches in the United States, the European Union, China, and India; public-private governance mechanisms; regulatory capacity in low- and middle-income countries; and the special problem of food additives. With this comprehensive approach, this book seeks to make significant progress in our understanding of global governance over critical consumer products sectors.

THE RISE OF A GLOBAL SYSTEM FOR FOOD AND PHARMACEUTICALS

The global scale of production, trade, and marketing of food and medicines has leaped since the Uruguay Round of trade liberalization agreements and the establishment of the WTO in 1995.[4] During the same period, bilateral, regional, and other multilateral trade and investment initiatives opened borders to greater participation not only from firms based in Canada, Europe, Japan, and the United States but also to food and medicines manufacturers based in Brazil, China, India, Mexico, and others.

Even for countries not participating significantly in the production and export boom, their national food and pharmaceutical systems are nevertheless increasingly integrated into global supply networks pushing growing volumes of trade in food and pharmaceutical products. Between 1971 and 2001, the growth in food imports was remarkable in developing countries, which saw a rise of 115%, compared to developed countries at 45%.[5] Between the 1960s and the early 1980s, developing countries transitioned from net agricultural exporters to net agricultural importers.[6] Growing food imports into developing countries are also increasingly processed. Trade in processed food products accounts for some 66% of agricultural trade.[7]

Trends in global food markets are apparent in the pharmaceutical sector as well, although developed countries remain both the largest producers and consumers of pharmaceutical products. There is rapid growth in the market and research environment in emerging economies such as Brazil, China, and India, leading to a gradual migration of economic and research activities from Europe to these fast-growing markets. In 2013, the Brazilian and Chinese markets grew by 17% and 14%, respectively, compared to an average market growth of 1% for the five major European markets and 3% for the U.S. market.[8] More importantly for the purposes of this analysis, the global pharmaceutical supply chain has added multiple links so that, for example, a U.S. finished drug may be produced using an active pharmaceutical ingredient (API) made in China and ingredients made in Europe, Japan, or the United States. These components may be shipped to India where the finished drug is manufactured and then imported into the United States for distribution. Packaging and repackaging may take place in other countries.

4. Buttel, F., McMichael, P., eds., 2005. New Directions in the Sociology of Global Development, 265-99. Emerald Group Publishing, Ltd; Petryna, A., Lakoff, A., Kleinman, A., 2006. Global Pharmaceuticals: Ethics, Markets, Practices, 218-19. Duke University Press.

5. Food and Agricultural Organization, 2004. The State of Agricultural Commodity Markets 2004, 16. Food and Agriculture Organization, Rome.

6. *Id.* at 14.

7. *Id.* at 26.

8. European Federation of Pharmaceutical Industries and Associations, 2014. The Pharmaceutical Industry in Figures. Available at http://www.efpia.eu/uploads/Figures_2014_Final.pdf.

Aggregate statistics of global pharmaceutical trade mask the extent to which value attributed to a finished product is added or captured elsewhere.

Large, complex, global corporations have been central players in the global integration of the food and pharmaceutical systems. They have extended their operations, both directly through top-down directives and indirectly through contracted services, vertically and horizontally, to the point that "national" food and drug systems are always subject to an important source of supranational or extranational decision making, which, in turn, has facilitated supranational or transnational regulatory structures.

For both food and pharmaceuticals, at least one step in the chain from manufacturer or grower to consumer—from production, trade, processing, and packaging (and, for food, retailing)—is typically managed by a global food or pharmaceutical firm.[9] Moreover, both industries are highly concentrated—and becoming more so—at the global level. According to the FAO, "a small number of companies now dominate each part of the food chain in OECD countries."[10] This trend is expanding to developing countries where global supermarket players are entering more low- and middle-income markets, shaping procurement and distribution systems.[11] The top 10 food retailers control 24% of the estimated US$3.5 trillion global food market while the top 10 food and beverage processing companies control 24% of the estimated US$1.25 trillion global market for packaged foods.[12]

The 10 largest drugs companies control over one-third of the global pharmaceuticals market, several with sales of more than US$10 billion a year and profit margins of about 30%. Six are based in the United States and four in Europe.[13]

The globalization of food and drug markets has produced important benefits, like increased access to medicines, diversifying varieties of foods available to consumers, new markets for producers, and growth opportunities for low- and middle-income countries that invest in specialized aspects of food and drug supply chains. China supplies about 43% of the world's APIs for anti-infective medicines and exports 77% of the active ingredients made in the country, a $4.4 billion industry. India exports 75% of the $2 billion worth of the APIs it produces. But globalization of food and drug markets has also negatively affected the integrity of critical supply chains and, in startling episodes, jeopardized consumer health and welfare. Contaminated food and drugs originating from poorly regulated facilities have caused thousands of deaths and hundreds of thousands of injuries. Falsified and substandard medicines not only sicken

9. Clapp, J., Fuchs, D., 2009. Corporate Power in Global Agrifood Governance, 5. The MIT Press.
10. Food and Agriculture Organization, 2003. Trade Reforms and Food Security: Conceptualizing the Linkages, 119. Food and Agriculture Organization, Rome.
11. Food and Agriculture Organization, 2005. State of Food and Agriculture 2005: Agriculture Trade and Poverty: Can Trade Work for the Poor? 21-22. Food and Agriculture Organization, Rome.
12. Clapp, J., Fuchs, D., 2009. Corporate Power in Global Agrifood Governance, 6. The MIT Press.
13. World Health Organization. Pharmaceutical Industry. Available at http://www.who.int/trade/glossary/story073/en/.

immediate recipients but also expand microbial resistance to affordable antibiotics, jeopardizing global population health. These episodes have been driving forces behind international and transnational initiatives to establish rules, standards, norms, and institutions to govern the global food and medicines systems. Global governance rules and institutions endeavor to create quality control mechanisms that ensure each of the steps in the food and drug chain is carried out in a manner that mitigates risks and maximizes benefits of transnational commercial activity. In the United States, for example, sweeping changes to food laws been embodied in the 2011 Food Safety Modernization Act aimed at international cooperation, overseas monitoring of food production facilities, and practices to effectively regulate importers.

CONCEPTUALIZING FOOD AND DRUG REGULATION IN GLOBALIZED MARKETS: INTERDISCIPLINARITY AND GOVERNANCE

It has been a standard feature of volumes addressing global regulation generally and food and drug regulation specifically to situate those volumes within the secondary literature of one or more academic disciplines or to make efforts toward constructing a unifying or coherent "framework" by which to analyze the subject. So Walter Mattli and Ngaire Woods situated their volume between "capture" and "public (or common) interest" theories of regulation with the editors attempting to assess which factors play the most important role in facilitating the latter type of regulatory regime. In *Corporate Power in Global Agrifood Governance*, Jennifer Clapp and Doris Fuchs synthesized the many aspects of transnational corporate influence through a framework divided into instrumentalist, structuralist, and discursive power. These features almost always serve as weaknesses either because the asserted cohesiveness is partially or wholly illusory or the commitment to cohesion sacrifices important, relevant perspectives. Reviewing Mattli's and Woods's *The Politics of Global Regulation*, Joel Campbell noted that "the chapter on human rights sticks out like a sore thumb among the finance and business-related pieces . . . a contribution or two by non-academic regulatory practitioners (either regulators or their corporate clientele) could have provided thought-provoking counterpoint."[14] Philip Howard noted of Clapp's and Fuchs's *Corporate Power* that it provided an "unsatisfactory unifying framework" and that "two of the chapters . . . challenge this approach and instead employ their own alternatives."[15]

A second common feature of edited volumes committed to globalization (whether broadly or topic specific) is the tendency of the introductory chapter to summarize contributors' arguments and provide relatively succinct biographical notes in a separate compendium, often at the end of the volume. To some extent

14. Campbell, J., 2010. Book Review. International Affairs 86(2) 2010, 561-563.
15. Howard, P. K., 2010. Book Review. American Journal of Sociology 115(6) 2010, 1915-1916.

this should be uncontroversial: all the contributors would be presumed to be experts on their subject and have proceeded with their analysis according to the appropriate methodology of their discipline, regardless of institutional affiliation, past, or current experience. Yet scholars of anthropology, cultural studies, history, literary criticism, and other social science disciplines have long known that authors and their texts are not neatly separated and that the meaning of the text is often intricately tied to the place and context from which the researcher observes, analyzes, collaborates, and writes.[16]

This volume therefore differs from previous studies in both its approach to interdisciplinarity and the explicitness of the link between contributors, their backgrounds, and their observations on regulatory gaps, joint public-private approaches to oversight of expanding food and drug supply chains, and the pressures that face globally competitive (if increasingly concentrated) industries, nation-states that desire to both support strong industries while regulating their conduct, and consumers, who face the most substantial risks of injury and death from poorly or inadequately regulated products that originate in whole or in part from growing and diversifying sources. "Interdisciplinarity" in its current rendering overwhelmingly refers to the application of one or more academic disciplines like epidemiology and economics or history and engineering.[17] Even scholars who take a more comprehensive view of interdisciplinarity like Robert Frodeman, Julie Thompson Klein, and Carl Mitcham devote the substantial bulk of their work toward interdisciplinarity in academic disciplines.[18] This volume, by contrast, adopts a mode of interdisciplinarity that takes as its focus supply chains for food and drugs that now stretch across the globe and, for any given jurisdiction, implicate questions related to the rights, roles, and responsibilities of firms attempting to maximize profits while obeying either or both of the letter and spirit of the law; consumers making choices based on basic levels of trust placed in brands built by firms, certifications provided through nongovernmental third parties, or regulators; regulators endeavoring to apply the most efficient forms of oversight to conserve scarce resources; and, of course, scholars of anthropology, consumer economics and microeconomics, history, international relations, law, medicine, nutrition science, political science, public health, and regulatory science.

Taking food and drug supply chains, including relevant regulatory checks along supply chains' routes, as a focus without more might be regarded as another species of "multidisciplinarity"; that is, juxtaposing perspectives on a shared topic or problem in serial order.[19] That undertaking alone would be worth the effort. Understanding the complexity and changes to food and drug

---

16. Said, E., 1983. The World, the Text, and the Critic. Harvard University Press.
17. University College London. More About Interdisciplinarity. Available at http://www.ucl.ac.uk/basc/faq/interdisciplinarity.
18. Frodeman, R., Klein, J. T., Mitcham, C., 2012. Oxford Handbook of Interdisciplinarity. Oxford University Press.
19. Frodeman, R., Klein, J. T., Mitcham, C., 2012. Oxford Handbook of Interdisciplinarity, 17. Oxford University Press.

markets wrought by globalization would require exploring the perspectives of not only those emphasized in this volume but also small and large grocers, independent and chain pharmacists, specialists in global anticorruption laws, and hundreds of others. But the contributions in this volume have been developed after an iterative process in which contributors collaborated with each other a priori in a multidisciplinary way, then adapted their contributions in light of those collaborative conversations. So, for example, Jesse Goodman's chapter on emerging challenges in the pharmaceutical development ecosystem, written from the perspective of an academic and regulator, was developed after several conversations with Sandra Eskin, who analyzes similar global challenges in the food sector, but from the perspective of a long-time consumer advocate. The chapter contributed by Sam Halabi and John Monahan represents a collaborative effort between a policymaker and a legal scholar studying the state of pharmaceutical regulatory capacity in low- and middle-income countries as observed during the 2009 H1N1 pandemic influenza vaccine distribution experience.

This is not to overstate either the uniformity in approach by contributors to this volume nor its contribution to an understanding of interdisciplinarity generally. This background is meant only to clarify that in addition to juxtaposing and coordinating voices from different managers of, and participants in, globalized food and drug supply chains, those voices have interacted, blended, and, to various degrees, integrated in the chapters contained herein. Indeed, U.S. Food and Drug Administration Commissioner Margaret Hamburg's Foreword to this volume embodies the kind of interdisciplinarity shaping the behavior of firms, consumers, and regulators alike. The FDA is working with think tanks, examining models from aviation and finance, and considering a wide range of public-private partnerships as it shapes its role as a global agency. In contrast to current approaches to the challenges facing both national and international food and drug regulators, the volume as a whole contributes a heretofore lacking interdisciplinary voice.

Relatedly, the volume does not view the individual contributions and the volume as a whole as more relevant than, or superior to, the relationship between the authors and their texts. This introduction itself includes far more information about contributors' backgrounds than current treatments, as does the section of the volume dedicated specifically to contributor biographies. To some extent, this aspect of *Food and Drug Regulation in an Era of Globalized Markets* does what other disciplines have long argued or at least implied: the context and position of the scholar/participant matter and the reader should assess the text with an eye toward that context. Consistent with the view of interdisciplinarity taken generally, it is essential to understand the perspective contributors adopt when assessing problems and solutions in global supply chains for food and drugs. Attorneys for pharmaceutical firms, for example, may view globalization as posing difficulties in navigating differing national regulatory systems for each stage of drug development, especially from clinical trials to Phase 4 postmarketing surveillance. Consumers of increasingly imported food products, by contrast, are concerned with what information regulators have—and how they use it—about the farms, fisheries, and facilities from which foreign food originates and through which it is handled.

The latter characteristic of this book is consistent with what many scholars of regulated industries now acknowledge to be the shift from "regulation"—action by the state or, at the international level, by groups of states, to control the conduct of economic actors through mandatory legal rules with monitoring and coercive enforcement—to governance, a term that in this context means

(1) incorporat[ing] a decentralized range of actors and institutions, both public and private, into the regulatory system, as by negotiating standards with firms, encouraging and supervising self-regulation, or sponsoring voluntary management systems;
(2) rely[ing] on this range of actors for regulatory expertise;
(3) modify[ing] its regulatory responsibilities to emphasize orchestration of public and private actors and institutions rather than direct promulgation and enforcement of rules; and
(4) utiliz[ing] "soft law" to complement or substitute for mandatory "hard law."[20]

In other words, regulation as the coercive power of the state has given way to collaborative approaches to industry behavior informed by a broad range of stakeholders from small producers to complex, global firms, to consumers and patients.[21] It is those stakeholder views, augmented through a mutually iterative exchange, which this volume endeavors to capture and convey.

THE PLAN OF THIS BOOK

The views presented in this book document the perspectives of attorneys, consumers, firms, governments, and scholars who today shape both changing national regulatory regimes as well as global food and drug governance. The various chapters investigate a large range of activities by corporate actors; the linkages between a corporation's global activities and the national and international regulatory regimes shaping those activities; the interaction between stakeholders; as well as the proliferating types of governance mechanisms such as public-private partnerships, firm-driven supplier standards, borrowing of regulatory approaches from other economic sectors, and joint public health emergency planning between citizens, firms, and governments.

---

20. Abbott, K., Snidal, D., 2009. Strengthening International Regulation through Transnational New Governance: Overcoming the Orchestration Deficit. Vanderbilt Journal of Transnational Law 42, 509.
21. Strange, S., 1996. The Retreat of the State: The Diffusion of Power in the World Economy, 4. Cambridge University Press.

The book is divided into two parts. Part I provides insights into the governance, regulation, and most critical vulnerabilities in globalized pharmaceutical supply chains. Part II then focuses on national and international regulatory structures, including entirely firm-driven ones, which are emerging to manage the diversifying and globalizing sources of food and food additives. The chapters in both sections contribute unique insights on regulatory problems in global food and drug governance and implications for debates over the future of assuring safety and quality in the food and drug supply.

In Chapter 1, Jesse L. Goodman, former chief scientist at the U.S. Food and Drug Administration and current director of Georgetown University's Center on Medical Product Access, Safety, and Stewardship provides an overview of the critical challenges facing the global pharmaceutical development system as well as the private-public partnerships that will facilitate wider access to better medicines and ensure the supply of essential, existing ones. Starting from the recognition that drug development costs have soared and regulatory requirements have increased, Goodman explains the ways that clinical trial practices may be adapted to facilitate development and approval processes for new drugs, link economic and regulatory policies to keep the drug supply safe, and assemble teams of "big data" analysts, physicians, and regulators to ensure that growing aggregate-data technologies are effectively harnessed and used to develop and target therapies. Specifically, he examines the role that adaptive trial design, multiregional clinical trials, and new approaches to clinical endpoints may play in opening possibilities for facilitating drug development. He then outlines the most important vulnerabilities in the pharmaceutical supply chain after drugs have been approved, explains why those vulnerabilities have emerged, and explores the regulatory harmonization and regulatory investment approaches that will be necessary to address those vulnerabilities. Despite the importance of so-called big data capabilities to solving problems from drug development, to approval, to the prevention of falsified and counterfeit medicine operations, Goodman urges a carefully planned, team-based approach to the use of big data in order to avoid needless and misleading phenomena that often emerge from current applications, and to focus promising analytic capabilities toward the medical needs where they will provide the greatest benefit.

Marc Scheineson, the head of Alston & Bird's Food and Drug Law Practice, a former legal adviser to the Health Subcommittee of the U.S. House of Representative's Ways and Means Committee and former commissioner of legislative affairs for the U.S. Food and Drug Administration, provides detail to Goodman's overview of supply chain vulnerabilities in Chapter 2 by examining the changing nature of regulatory and enforcement activities undertaken by the U.S. FDA as it becomes a global consumer protection agency. He covers not only the legislative and administrative bases and priorities for FDA oversight, but how its investigation and enforcement activities have changed with its mandate to inspect foreign facilities. He argues that the emergence and growth of electronic registration laws together with expanded presence of inspectors overseas is ultimately shaping how the FDA views its priorities and its regulatory partnerships. His chapter highlights regulators' complex relationship with other actors, including firms, regulatory partners, and courts.

In Chapter 3, Sabine Haubenreisser, who serves as liaison from the European Medicines Agency (EMA) to the U.S. Food and Drug Administration and is therefore deeply involved in regulatory cooperation and harmonization activities, and EMA Principal International Affairs Officer Martin Harvey-Allchurch contrast the role and mission of the U.S. FDA from Scheineson's chapter to that of the EMA as it both coordinates European member state activity and participates in collaborative regulatory initiatives with national and international regulatory partners. They specifically address the means by which current regulatory cooperation efforts provide for more efficient use of resources in support of risk-based approaches, contribute to increased consistency and predictability in regulatory regimes, and support a global approach to development, authorization, and supervision of medicines.

In Chapter 4, Haas and Epstien, attorneys at the global law firm Latham & Watkins, elaborate upon the clinical trial issues Goodman raised in Chapter 1. Declining clinical trial volunteer and eligibility rates as well as cost in many developed countries are encouraging firms to (1) find jurisdictions with a larger number of eligible volunteers and trial participants and (2) transfer some or all of the legal obligations connected with clinical trials to contract research organizations (CROs). Haas and Epstien trace the growing and changing national and international regulatory framework aimed at both protecting human research subjects and ensuring that safety and efficacy data effectively and legally support drug approval applications. The authors provide not only detailed analysis of the current legal structures in place, but the practical aspects of deciding whether to conduct clinical trials in one or more foreign jurisdictions, the criteria by which CROs should be evaluated, and idiosyncrasies related to cultural and site-specific arrangements that may affect trial integrity.

In Chapter 5, Halabi and Gostin, who chaired the U.S. Institute of Medicine's analysis of falsified and substandard medicines in global supply chains, analyze the global markets for APIs, excipients in which those ingredients are suspended, and how the changing nature of those supply chains, which increasingly originate in or pass through China and India, are increasing vulnerabilities to the introduction of falsified and substandard medicines. Halabi and Gostin explore the global public health impact from these vulnerabilities, which include not only direct injury and death resulting from adulteration and contamination, but even more catastrophic damage in the form of global microbial resistance to false or diluted antibiotics. The authors explain current barriers to solving the problem of falsified and substandard medicines and recommend solutions that favor commitment of resources to areas in which the public health burden is heaviest: low- and middle-income countries.

In Chapter 6, Halabi and Monahan provide more detail as to the topic of investment in regulatory capacity in low- and middle-income countries outlined by Commissioner Hamburg in the Foreword and Goodman in Chapter 1. They use the global response to the 2009 H1N1 influenza pandemic to analyze the capacity of national regulatory authorities to approve medicines generally and to react to public health emergencies of international concern specifically. Bringing perspectives from international law and international politics, and health diplomacy, the authors address three aspects of low- and middle-income countries' regulatory capacity: (1) the relationship between regulatory barriers and legal barriers that caused pharmaceutical companies to delay participation in the global response; (2) the lack of regulatory capacity in low- and middle-income countries that hindered efforts to distribute pandemic vaccine even after legal agreements with pharmaceutical companies and governments were in place; and (3) the role that investment in regulatory capacity in the future could play in establishing timely, harmonized, and efficient procedures for facilitating vaccine and antiviral medical countermeasures during public health emergencies. Their chapter brings into particular focus the critical nexus between international regulatory authorities based at the World Health Organization, global pharmaceutical firms positioned at the center of global access to medicines, and national regulatory authorities in low- and middle-income countries who struggle to effectively engage with the latter two stakeholders because of inadequate resources that may be devoted to regulatory capacity and personnel.

In Chapters 7 and 8, Suneja and Schatz elaborate upon changes in the regulatory environments in India and China, respectively, that have both facilitated and responded to growth in their pharmaceutical sectors and the context in which those changes should be understood. Suneja, a former first secretary for Market Access & Compliance Attaché at the U.S. Embassy in New Delhi who now advises private sector clients on Indian pharmaceutical regulation, provides an assessment of each ministry in the Indian government that regulates aspects of the pharmaceutical sector, the effect that the election of BJP leader Narendra Modi has had upon those ministries, as well as wider legislative initiatives that seek to balance consumer, patient, innovative, and generic pharmaceutical firm interests. Schatz, an attorney with experience advising clients from both China and the United States, provides a practical, insider view on regulatory priorities at the China Food and Drug Administration and how those priorities have been shaped by the national government's wider economic strategy. His contribution not only details specific regulatory changes but also aspects of the Chinese legal environment that have an increasingly important effect on the pharmaceutical market, such as the need to redistribute growth from the major coastal cities to the interior and the ways in which anticorruption laws are being used to influence the behavior of both foreign and domestic players.

Part II begins with Derek Yach's, Eleanor Alexander's, and Adriana Selwyn's analysis in Chapter 9 of the fundamental issues surrounding any governance regime for globalized food markets: sustainability, nutritional content, and public health. Contrasting global food governance with global tobacco governance—which is embodied in a binding international treaty—Yach, a former senior vice-president of Global Health and Agriculture Policy at PepsiCo and executive director for Noncommunicable Diseases and Mental Health at the World Health Organization, and his coauthors argue that while tobacco and even alcoholic beverages may be taxed and regulated with some level of product specificity and coherence, doing so for food and nonalcoholic beverages is not similarly practicable. Therefore, public-private partnerships are an essential aspect of balancing firm, consumer, and public health interests. The authors outline the circumstances under which these partnerships are most likely to create win-win scenarios for regulators and firms, and how they may be designed to avoid both the appearance of and actual regulatory capture.

In Chapter 10, Lisa Heinzerling, William J. Brennan, Jr. Professor of Law at Georgetown University Law Center and a former associate administrator at the U.S. Environmental Protection Agency, analyzes the regulatory structure of food under the U.S. system. She not only points out the gaps, inconsistencies, and inefficiencies arising from fractured agency regulation of food—a lesson applicable to a large number of jurisdictions worldwide—but how fracturing may be used to manipulate, vet, and orchestrate regulations to fit a political rather than public health agenda. Her analysis contains important lessons for regulatory design not only for states now considering food regulatory system design, but also for advocates of reform for the U.S. system specifically.

In Chapter 11, Halabi examines the growing partnerships and international agreements that the U.S. FDA has put in place with national regulatory authority partners and situates those agreements within the emerging literature on global administrative law. He points out that the complexity of the regulatory environment has encouraged the FDA to take a broad approach to both product categories as well as applicable standards. Halabi provides a taxonomy of FDA agreements, which range from relatively "soft" confidentiality commitments to "harder" agreements recognizing entire food safety systems and committing agencies to concrete cooperative action. He argues that the complexity and diversity of the FDA's agreements carry important implications for the debate about the emergence and strength of global administrative law as that term is now understood.

In Chapter 12, Eskin, the director of the Pew Charitable Trust's food safety project and a member of the Food Safety Modernization Act Surveillance Working Group at the Centers for Disease Control and Prevention, addresses the role of food additives in the current food production system and how those additives may be effectively regulated. Generally, food additives are used by manufacturers and processors to improve and maintain the nutritional value of food, texture, safety of packaged foods, taste, and appearance, including color. Changes in the global market for food additives reflect the broader trends identified throughout the book. While the United States and Europe continue to serve as the largest producers and consumers of food additives because of fat replacers, artificial sweeteners, and hydrocolloids as well as additives that support a convenience diet, many leading manufacturers of food additives have set up production bases in emerging economies and targeted growing demand in those countries. China, for example, has become a leading producer and exporter of certain varieties of food additives, especially vitamins. While Eskin's analysis of food additive regulation is specific to the U.S. experience, the three core principles outlined in the chapter—independent review, effective testing, and comprehensive health assessment—provide a useful model for any system endeavoring to effectively regulate food additives, as those ingredients constitute an ever-growing portion of global food consumption.

In Chapter 13, Waldrop, the director of the Consumer Federation of America's Food Policy Institute, analyzes three key aspects of the 2011 Food Safety Modernization Act (FSMA), the most sweeping and comprehensive update to U.S. food laws in 70 years that affect consumers.[22] The law's intent is to prevent problems that can cause foodborne illness and enable the FDA to have a record of facilities processing food for sale in the United States, a mandate that expands the FDA's already global regulatory activities. The FSMA gives the FDA broad new powers to prevent food safety problems, detect and respond to food safety issues, and improve the safety of imported foods.[23] The law specifically aimed to update FDA authority in light of the reality of global food and food additive markets.[24] Waldrop examines the scope and depth of the information the FDA needs, according to consumers, to assess the safety and reliability of all facilities, including overseas facilities. He also provides a separate analysis of the Foreign Supplier Verification Program.[25] The verification component of the program is essential in order to provide importers with assurances that their supplier is meeting its obligations to provide safe food. Consumers groups believe that annual on-site verification of an importer's suppliers is critical as an on-site review can provide much greater assurance than a paperwork check.[26]

In Chapter 14, Glasner analyzes the transitions in international investment and trade law that have influenced the regulatory standard-setting role of the Codex Alimentarius Commission, an entity formed by, and nominally still is, a collaboration between the UN agencies Food and Agricultural Organization and the World Health Organization. Glasner investigates the twin, sometimes

22. Food and Drug Administration. FDA Food Safety Modernization Act (FSMA). Available at http://www.fda.gov/Food/GuidanceRegulation/FSMA/.

23. Agency of Agriculture, Food, and Markets. Food Safety Modernization Act. Vermont.gov. Available at http://agriculture.vermont.gov/food_safety_consumer_protection/fsma.

24. Department of Health and Human Services, Food and Drug Administration, 2013. FDA's International Food Safety Capacity-Building Plan. Available at http://www.fda.gov/downloads/Food/GuidanceRegulation/UCM341440.pdf.

25. Food and Drug Administration, 2013. Foreign Supplier Verification Programs for Importers of Food for Humans and Animals. Federal Register Vol. 78, No. 145, 45,730-45,779.

26. Food and Drug Administration, 2014. Foreign Supplier Verification Programs for Importers of Food for Humans and Animals. Federal Register Vol. 79, No. 188, 58,574-58,599.

competing missions of Codex to issue standards, guidelines, and codes of practice that protect consumers and promote fair trade. She shows that the prominence given Codex in the World Trade Organization's Agreement on Sanitary and Phytosanitary Measures has drawn both greater numbers of corporate participants as well as less transparent procedures. Using some disputes over international standards, she demonstrates that international organizations committed to regulatory standard-setting roles in the food context must undertake their responsibilities transparently and fairly, or risk undermining standards that may facilitate access to affordable, nutritious food moving in international commerce.

In Chapter 15, Lin pivots to the "private" side of the public-private partnerships with which Yach and his coauthors framed the food regulatory issue. He shows that standards and policies developed entirely by private firms (although perhaps in the shadow of national and international laws) are just as important, if not more so, than statutes or regulations that purport to ensure food safety and how those private regimes are providing both costs and benefits to consumers. Indeed, his chapter provides additional context to Glasner's as Codex is increasingly becoming a global public-private partnership organization in nature, even though its technical structure is that of a forum for state regulatory authorities.

ACKNOWLEDGMENTS

The author would like to thank the O'Neill Institute for National and Global Health Law at Georgetown University for funding the workshop that facilitated this project. Special thanks go to Institute directors and staff who made the project possible: Oscar Cabrera, Aliza Glasner, Caroline Gould, Larry Gostin, Nan Hunter, Susan Kim, John Monahan, Joe Page, Phoebe Ramsey, Mike Templeton, Alison Woodworth, Susan Weld, Tim Westmoreland, and David Vladeck. I would also like to thank three anonymous reviewers for helpful comments.

Acknowledgments

This book originated with a week-long program at the O'Neill Institute for National and Global Health Law at Georgetown University, which brought together academic researchers, consumer advocates, attorney practitioners, regulators from every continent, representatives from pharmacopeial organizations, and students to explore the theme of emerging issues in food and drug regulation. That program would not have been possible without the generous contributions of time and energy from representatives from the U.S. Food and Drug Administration's Office of Global Regulatory Operations and Policy, Center for Drug Evaluation and Research, and the Office of Foods and Veterinary Medicine with thanks to Howard Sklamberg, Mary Lou Valdez, Katherine Bond, Jennifer Devine, Ilisa Bernstein, Sharon Mayl, Doug Stearn, and particular gratitude to Katherine Cooper in the Office of Strategy, Partnerships, and Analytics within the Office of International Programs. I would like to thank the O'Neill Institute for National and Global Health Law, Georgetown University Law Center, and the O'Neill Institute Faculty for financial support for that workshop. I am grateful to Nancy Maragioglio, Billie-Jean Fernandez, Chris Wortley, and Carrie Bolger at Elsevier for shepherding the project through the publication process, and to Jenelle Beavers and three anonymous reviewers for providing helpful comments. I would also like to thank Fernanda Alonso, Brian Gibbons, Karen Silva Mora, Katy Spraberry, Claudia Verónica, Torres Patiño, and Emily Wong for research and editorial assistance. Finally, I would like to express my gratitude to the contributors to this book, as well as my friends and family, for their patience and support throughout the publication process.

Part I

Governance, Regulation, and Vulnerabilities of Globalized Pharmaceutical Supply Chains

Chapter 1

Addressing Emerging Challenges in the Pharmaceutical Product Development Ecosystem

Jesse Goodman*

*Center on Medical Product Access, Safety and Stewardship (COMPASS), Georgetown University Medical Center, Washington, District of Columbia, USA

Chapter Outline

INTRODUCTION

Drug and medical device regulation over the next several decades will be fundamentally shaped not only by the expansion of their respective supply chains—which has already introduced critical vulnerabilities covered by contributors to this volume—but also by advances in regulatory science that offer the potential to reduce barriers and enhance the efficiency of drug development, approval, and surveillance; harness new analytical methods and incorporate new areas of therapeutic potential. The following chapter analyzes such trends with a specific focus on the governance relationships that will characterize product development and distribution as product developers and providers (from upstream innovators and their financiers to downstream retailers, physicians, and pharmacists), regulators, consumers, and patients work together to manage the benefits and costs of

Food and Drug Regulation in an Era of Globalized Markets. http://dx.doi.org/10.1016/B978-0-12-802311-2.00001-1

innovation generally and try to protect what gains have been made from unintentional erosion by, for example, microbial resistance and vulnerabilities in supply chains. The chapter will discuss selected aspects of the costs and complexities of the pharmaceutical and medical device product development ecosystem, and whether it can become more efficient. It will also discuss vulnerabilities and potential solutions in the pharmaceutical and medical device ecosystems, ways to address antimicrobial resistance, and the opportunities and challenges of big data.

APPROACHES TO REDUCE DRUG DEVELOPMENT COSTS AND SPEED INNOVATION

Despite the expanding access to medicines facilitated by globalization, the costs involved in clinical development have increased exponentially over the past 30 years. The average cost to develop one new drug nearly doubled from 2000 to 2005 alone.[1] The time to develop one drug from an early candidate to an approved product often exceeds 10 years.[2] A brief overview of the drug development process in the United States will help understand why these costs are high and continue to escalate. The first phase of new drug development typically begins with identifying a candidate target for drug action followed by preclinical chemical synthesis to identify a family of molecules as candidate new chemical entities (NCE).[3] Basic preclinical research is then conducted to determine the candidate NCE's basic properties including safety in cell and animal models, and its pharmacodynamics and pharmacokinetic properties. These tests attempt to predict the NCE's effects in humans so that a lead candidate and a safe starting dose for human trials may be defined. An additional component of the preclinical research phase is to test the drug both *in vitro* and on relevant animal species for pharmacological activity and toxicity.

At the conclusion of the needed *in vitro* and animal studies, the sponsor may file an Investigational New Drug (IND) application with the FDA, at which point it becomes "a new drug subject to specific requirements of the drug regulatory system."[4] The FDA requires the IND to contain information and research analysis that pertains to three general inquiries: pharmacology and toxicology studies, a detailed report on the properties of and manufacturing process for the drug, and a proposed protocol for human trials which includes safeguards to ensure that human test subjects are not exposed to any unnecessary risks. The FDA's IND review process is primarily designed to ensure that the drug company's proposed protocol conforms to regulatory safeguards and that the trials

1. Drugs.com. New Drug Approval Process. Retrieved from http://www.drugs.com/fda-approval-process.html.
2. *Id.*
3. U.S. Food and Drug Administration, 2014. Investigational New Drug Application. Retrieved from http://www.fda.gov/Drugs/DevelopmentApprovalProcess/HowDrugsareDevelopedandApproved/ApprovalApplications/InvestigationalNewDrugINDApplication/default.htm.
4. *Id.*

will not expose human subjects to unacceptable risk and also that the studies are likely to produce useful information.

Once an IND is allowed to go forward, the sponsor typically embarks on a three-phase clinical development process intended to lead to approval. Phase 1 typically focuses on safety and may involve 20-80 human subjects to detect any major safety issues and to evaluate the drug's distribution, metabolism and excretion.[5] Phase 2 trials typically involve several hundred individuals and focus on the dose-response to the candidate drug. They are designed to identify the most appropriate dose or doses for further studies, as well as to identify safety issues that may not have been apparent in the smaller number of subjects included in Phase 1. A control group is often included and administered a placebo while the test group receives the drug.[6] Phase 3 studies are typically referred to as "pivotal" and are designed to establish the efficacy and safety of the product, and support its approval, for the intended indication(s) and in the intended population(s). They are typically randomized and well-controlled studies comparing the drug against a placebo or standard of care and measuring effects on meaningful clinical endpoints. Phase 3 studies for common diseases typically involve thousands of patients and, as such, also offer an important opportunity to identify less common adverse effects not seen in earlier studies.[7]

At the conclusion of clinical testing, if studies support a favorable benefit to risk profile, the drug manufacturer submits a New Drug Application (NDA) or, for a biologic drug, a Biologics Licensing Application (BLA) to the FDA. The NDA or BLA is extensive, often requiring a significant amount of time and money to produce; it includes comprehensive information on chemistry, manufacturing processes and specifications, clinical data supporting each dosage and dosing form/route the company intends to use, proposed packaging and labeling, the results of all relevant preclinical (e.g., laboratory and animal studies of drug effects, including toxicology) and clinical testing.[8] The NDA also typically provides plans for both safety surveillance, which may be needed to address any safety signals or questions identified during development, as well as for any planned Phase 4 studies be intended to be performed post-marketing. Phase 4 studies typically are performed to study unresolved questions from the clinical trials, for example potential safety signals, as well as safety and effectiveness in additional populations or, for products which were approved through FDA's accelerated approval pathway, based on likely surrogate markers for efficacy, effects on actual clinical endpoints. The complete application is reviewed by the FDA's Center for Drug Evaluation and Research (CDER) or its Center for

5. U.S. Food and Drug Administration, 2014. Drug Approval Process. Retrieved from http://www.fda.gov/Drugs/ResourcesForYou/Consumers/ucm295473.htm.

6. *Id.*

7. *Id.*

8. U.S. Food and Drug Administration, 2014. How Drugs are Developed and Approved. Retrieved from http://www.fda.gov/Drugs/DevelopmentApprovalProcess/HowDrugsareDevelopedandApproved/ApprovalApplications/NewDrugApplicationNDA/default.htm.

Biologics Evaluation and Review (CBER) by a multidisciplinary review team that typically includes biologists, physicians, statisticians and epidemiologists, chemists, manufacturing experts, pharmacologists, and other scientists. The team must evaluate whether the drug is safe, effective, and whether its proposed benefits outweigh its potential risks. Additionally, the team must review whether the proposed manufacturing process is well controlled, consistent and high quality in order to maintain the drug's safety and quality, and whether the proposed product specifications and labeling are appropriate. Finally, if not already done, the FDA conducts an inspection of the manufacturer's facilities to ensure they conform to regulatory guidelines. Upon final approval, the manufacturer is free to bring the drug to market. However, the manufacturer must continue to submit safety and manufacturing updates to the FDA to ensure that the product performs as expected and must also complete and submit the results of any Phase 4 studies.

There are various approaches that can be considered to increasing efficiency and reducing the very high costs of medical product development. First, regulatory science can create better tools and methods for product evaluation and reduce uncertainties both for new and emerging technologies as well as to improve preclinical prediction and clinical evaluation of safety and efficacy. Second, new technologies and alliances can increase speed and efficiency of clinical trials and development. Third, enhanced postapproval capacity to detect uncommon or unexpected adverse events, such as through active monitoring of large populations using electronic health record, can enable faster detection of potential problems after approval and thus support, where appropriate, earlier marketing of products that truly address unmet medical needs based, potentially, on smaller pre-approval safety databases. Just approving a product does not ensure access, even for a potentially lifesaving innovation. To help speed access to important new products, in the United States, the FDA, is working with the primary healthcare cost reimbursement agency, the Centers for Medicare and Medicaid Services, to develop programs that help ensure development and access to the information needed to support and tie speed reimbursement for products addressing truly unmet needs and that provide clinically significant treatment effects.

Clinical Trials: Reducing Costs, Speeding Development: New Approaches to Patient Recruitment, Study Design, and Settings

Clinical trials required as part of drug approval are often, the most expensive aspect of bringing new drugs to market. The following approaches could enhance the efficiency or reduce costs in clinical trials: (1) deploying new technologies and/or social capital for patient recruitment; (2) adaptive/iterative study designs; and (3) identifying circumstances under which bigger or smaller populations may better serve scientific and regulatory needs. For example, highly effective products, such as those that are genetically targeted in so called "precision medicine," may need only small, well-designed studies to show efficacy, while in some settings such as in common, chronic disorders, large but simple trials may be more economical, and faster and yet have enhanced power to detect smaller change in endpoints, which is critical for chronic illness. Many late phase studies may and should be simpler; that is, focus on the most important safety and efficacy endpoints and measure fewer other factors less often.

Adaptive Design

In traditional clinical trials, factors like the study population, primary endpoint, clinically meaningful treatment difference, and measure of variability are fully and finally prespecified during planning in order to design the study.[9] Investigators then collect all data and perform analyses. Researchers' initial assumptions determine the design and can have major effects on the success of the study. Adaptive or iterative design allows greater flexibility between initial assumptions and clinical endpoints. There are many variations of adaptive or iterative design possible, but the basic concept is to rapidly incorporate certain types of results (clinical, biomarker, imaging, etc.) during the study in order to potentially inform or modify the design, helping to identify and target winning strategies/products, reject failures, and match responses and enrollment to specific patient marker/drug pairs, and select doses. A simple example, now included as part of almost all randomized controlled trials, is monitoring and potentially stopping a study if either futility or clear benefit is seen based on one or more interim analyses. This requires pre-specified and appropriate statistical approaches and corrections for what are essentially multiple analyses. Adaptive design "incorporates prospectively planned opportunities for modification of one or more specified aspects of the study design and integrates hypotheses based on analysis of accumulating data while the study is ongoing."[10]

When used appropriately, adaptive or iterative design studies can reduce the sample size and lead researchers more efficiently to answering the relevant clinical questions. For example, John Kairalla and his coauthors detail how enrichment adaptive design may allow researchers in a single study to winnow populations until they effectively match patients with specific genetic markers to optimal treatment courses:

> *Enrichment of a study population refers to ensuring that participants in a trial are likely to demonstrate an effect from treatment, if one exists. For example, there is benefit to enrolling participants lacking comorbidities, with a risk factor of interest (such as high blood pressure), and likely to be compliant. An extension known as adaptive enrichment designs fulfills the desire to target therapies to patients who can benefit the most from the treatment. In such designs, a trial initially considers a broad population. The first study period reveals participant groups most likely to benefit from the test agent (discovery phase). Subgroup members are then*

9. Kairalla, J. A., Coffey, C. S., Thomann, M. A., Muller, K. E., 2012. Adaptive trial designs: a review of barriers and opportunities. Trials 13, 145.
10. Alexander, B., Wen, P. et al. 2013. Biomarker-Based Adaptive Trials for Patients with Glioblastoma—Lessons from I-SPY2. Neuro-Oncology 2013 Aug;15(8), 972-978.

randomized to receive either the active agent or control (validation phase). Power for the chosen subgroups is increased due to the increased sample size in the subgroups, while non-promising groups are discarded. Adaptive enrichment designs have been praised for their ability to identify patient groups and undiluted effect sizes that can aid in the design and efficiency of replication studies. An appealing area for adaptive enrichment is pharmacogenetic research where it could allow for isolation of the one or two genetic marker subgroups that are predictive for treatment response. The approach can increase efficiency when identifiable genetic subgroups have increased treatment benefit.[11]

Clinical Trials: Globalization

Historically, clinical trials were planned and undertaken primarily in the developed economies where drugs would then be largely marketed. Over the last 2 decades, sponsors have increasingly moved clinical trials outside of the most developed economies for interdependent reasons related to cost, access to potential study participants, increasing product use and pharmaceutical markets in developing countries, shorter time required to obtain results, and the desire to better understand regional or population variations in responses. Multiregional clinical trials (MRCTs) are now common for both costly innovator drugs and for treatments aimed at diseases that may disproportionately affect low- and middle-income countries. Since 2002, the number of FDA-regulated investigators based outside the United States has grown by 15% annually while U.S.-based investigators have declined by approximately 5.5%.[12] Seth Glickman and his collaborators estimated in a 2009 article that one-third of all Phase 3 trials being run by the largest 20 pharmaceutical companies were being conducted solely outside the United States.[13]

Multiregional trials can, if done well, speed up recruitment and lower costs. If well designed and conducted, they can allow analysis according to global region which may identify important differences in treatment response, including safety and efficacy. MRCTs can also be very useful in studying therapies that address medical problems that particularly affect countries, regions or populations outside of the most developed nations. In a study of 17,826 patients with acute malaria who could not take medication by mouth and other antimalarial treatments were not available, an MRCT with protocols in Bangladesh, Ghana, and Tanzania demonstrated the benefit of an artesunate rectal suppository in

11. Kairalla, J. A., Coffey, C. S., Thomann, M. A., Muller, K. E., 2012. Adaptive trial designs: a review of barriers and opportunities. Trials 13, 145.
12. Getz, K. A., 2007. Global clinical trials activity in the details. Applied Clinical Trials. Retrieved from http://appliedclinicaltrialsonline.findpharma.com/appliedclinicaltrials/article/articleDetail.jsp?id=453243.
13. Glickman, S. W., McHutchison J. G., Peterson E. D., Cairns C., Harrington R. A., Califf, R. M., Schulman, K. A., 2009. Ethical and scientific implications of the globalization of clinical research. New England Journal of Medicine 360, 816-823.

multiple African populations.[14] Furthermore widespread mobile phone use can help enable both patient recruitment and follow-up, and, together with lower costs can now help make clinical research both more feasible and more economical in low- and middle-income countries as well.[15]

MRCTs are not a panacea for addressing drug development speed and costs, and require thoughtful design and conduct and appropriate and vigorous regulatory oversight across a range of factors. Ethical issues are important and include avoiding undue incentives and ensuring fully informed consent, participation and the monitoring and protection of study participants. As in all clinical studies, investigators must be fully qualified and reporting of study results timely, complete and accurate.

In planning, conducting and analyzing results of MRCTs there are special challenges due to differences seen in different geographic areas and populations in factors such as disease diagnosis and severity, comorbidities, access to and standards of care, over-the-counter medicine use, nonmedical care, host genetic and other pharmacologic and physiologic factors, nutrition, diet, environment, health literacy, research ethics, culture, economic well-being, family and support systems, perceptions, and protocol compliance. These variations may result in difficulty determining whether outcome differences among regions detected in a trial are real or random. The best approach to manage these variations is to design studies and analyses up front that reduce or account for such factors and that allow for their analysis when considering trial results.

New Approaches to Endpoints: Biomarkers, Patient Reporting, and Remote Monitoring

The objective of clinical research generally is articulated in terms of measuring the effects of a treatment on an established clinical endpoint. For serious disorders, such endpoints typically are major clinical outcomes such as survival, but may also include well established more proximate markers of response that have been shown to be predictive of longer term clinical outcomes, such as blood pressure response for antihypertensives or reconstitution of immune function for HIV drugs.

As with adaptive study design and MRCTs, there are many promising new approaches to identifying and validating endpoints that may further decrease the time it takes to obtain results and, therefore, help speed the development process, identifying promising treatments more quickly and, conversely, reducing costs and patient risks by rapidly identifying treatments unlikely to be effective. Biomarkers, imaging, and other more proximate predictors of clinical outcome can offer promise of more rapid, predictive assessments of safety and

14. http://www.sgul.ac.uk/media/news-archive/2008/anti-malaria-pioneers.

15. Prue, C. S., Shannon, K. L., Khyang, J., Edwards, L. J., Ahmed, S., Ram, M., Shields, T., Hossain, M. S., Glass, G. E., Nyunt, M. M., Sack, D. A., Sullivan Jr., D. J., Khan, W. A., 2013. Mobile phones improve case detection and management of malaria in rural Bangladesh. Malaria Journal 12, 1-7.

efficacy.[16] Intraocular pressure, visual fields, and structural measurements of the optic nerve, for example, may serve as candidate endpoints in glaucoma clinical trials.[17] In addition, novel technologies such as those that can continuously monitor and report subjects' heart rate and rhythm blood pressure, oxygenation, breathing, etc., often via smartphones, can both create economies and increase the richness and accuracy, and, for patient reported outcomes of symptoms and functional status, the meaningfulness of clinical trial data. Similarly, postmarket surveillance for efficacy and safety can be potentially facilitated by patient-reported or directly monitored events via smartphones, SMS, and interactive messaging.[18] Like any measurement, however, all new approaches must be validated.

Such advances in remote sensing and monitoring, and in developing and measuring patient reported outcomes, offer tremendous promise not only for clinical trials but for more routine management of diagnosed conditions.[19] Studies have demonstrated the potential for improvement in self-management of diabetes using text messages.[20] Monitoring stress through the skin, and monitoring gait for movement disorders are also new tools that can be used both to enhance studies and routine care.[21]

THE SUPPLY CHAIN: HOW ARE WE VULNERABLE AND WHAT ARE POTENTIAL SOLUTIONS?

The aforementioned innovations demonstrate that both globalization and advances in technology have opened possibilities for a range of new measures and approaches with the potential both to enhance the accuracy and meaningfulness of studies and reduce the costs of bringing new therapies to market. Globalization has also brought with it fierce competition in the production and distribution of all

16. Institute of Medicine, 2010. Evaluation of Biomarkers and Surrogate Endpoints in Chronic Disease. National Academies Press, Washington, D.C.

17. Medeiros, F. A., 2014. Biomarkers and surrogate endpoints in glaucoma clinical trials. British Journal of Ophthalmology 2014 May; 99(5):599-603.

18. Li, F., Harmer, P., Liu, Y., Eckstrom, E., Fitzgerald, K., Stock, R., Chou, L. S., 2014. A randomized controlled trial of patient-reported outcomes with tai chi exercise in Parkinson's disease. Movement Disorders 29, 539-545.

19. Reichel, A., Rietzsch, H., Ludwig, B., Rothig, K., Moritz, A., Bornstein, S. R., 2013. Journal of Diabetes Science and Technology 7, 156-162; Faurholt-Jepsen, M., Vinberg, M., Christensen, E. M., Frost, M., Bardram, J., Kessing, L. V., 2013. Daily Electronic Self-Monitoring of Subjective and Objective Symptoms in Bipolar Disorder—the MONARCA Trial Protocol (MONitoring, treAtment and pRediCtion of bipolAr disorder episodes): A Randomised Controlled Single-Blind Trial. BJM Open 3, 1-7.

20. De Jongh, T., Gurol-Urganci, I., Vodopivec-Jamsek, V., Atun, R., 2012. Mobile Phone Messaging for Facilitating Self-Management of Long-Term Illnesses. Cochrane Database of Systematic Reviews. Retrieved from http://www.ncbi.nlm.nih.gov/pubmed/23235644.

21. Cheng, Q., Juen, J., Li, Y., Prieto-Centurion, V., Krishnan, J. A., Schatz, B., 2013. GaitTrack: Health Monitoring of Body Motion from Spatio-Temporal Parameters of Simple Smart Phones. ACM Conference on Bioinformatics, Computational Biology and Biomedical Informatics, Health Informatics Symposium. Retrieved from http://healthit.illinois.edu/wp-content/uploads/2013/08/acmbcbhis2013_GaitTrack.pdf.

of the components of these therapies, lengthening supply chains and making them simultaneously complex, but, paradoxically, often also more convergent and dependent on limited numbers of suppliers, and thus more vulnerable to potentially fatal defects, whether due to quality lapses or to illicit production of counterfeit or substandard products or components. At least forty percent of drugs and 80% of active pharmaceutical ingredients destined for the U.S. market are now made elsewhere, especially China and India. Similarly, more than 50% of medical devices are now imported. A 2011 U.S. Department of Commerce study determined that three-quarters of the manufacturers of products, drugs, and devices deemed critical were found to depend on a non-U.S. source for at least one component, and a majority had no alternative supplier.[22] "Just-in-time" norms prevalent in global supply chains mean inventories without buffers for supply interruptions, and alternative suppliers are similarly unlikely to retain excess capacity.

It is, therefore, often increasingly difficult to know where product components may originate, what the producer is doing in another country, and whether all those contributing in the production of product ingredients, as well as importers have been subject to robust regulatory oversight and are complying with critical legal and regulatory requirements. The results of such complexities have engendered increased risks, and, in turn, included sentinel and recurrent episodes of poisonings, adulterated drugs, and drug shortages. In 2008, overly sulfated chondroitin, much cheaper than genuine heparin, was deliberately used to adulterate heparin, a common blood thinner, and killed approximately 81 people and injured hundreds of others. The same year, China reported an estimated 300,000 victims of melamine-contaminated milk, with six infants dying from kidney damage and tens of thousands of related hospitalizations. In both the US and abroad, lack of redundancy in supply chains for both API and finished products, as well as aging infrastructure and competition, which, for certain commoditized products, has been driven primarily by reducing costs, have led to severe shortages of many critical medicines, including saline, multiple important antibiotics, certain anesthetics and nitroglycerin. The consequences of such adulteration and shortage problems could become far worse.

One way to help keep the global supply chain safe is to enhance collaboration in ensuring robust regulation both at home and abroad and, including in all countries of origin, where needed backed by trade and economic policy. Solutions to global supply chain vulnerabilities will require building global regulatory capacity in supplier regions as well as implementing better science for testing and risk analytics including early detection of vulnerabilities, anomalies, and environmental events. There also needs to be an increase in clarity, awareness, and enhancement of responsibilities of manufacturers for their supply chain and suppliers. Enhanced technologies, such as a hand held counterfeit detector being studied by FDA, and enhanced manufacturing controls, can help, but, to keep people safe, worldwide, adulteration and quality problems must, largely, be prevented at their sources.

---

22. https://www.bis.doc.gov/index.php/forms-documents/doc_view/642-department-of-homeland-security-dhs-assessment-impact-of-foreign-sourcing-on-health-related-infra.

There are also relatively commonsense approaches to the problem of drug shortages that could be adopted as part of a larger strategy on core health infrastructure. Rewarding product quality and the reliability of supply through reimbursement would be a major step forward. In addition, we should consider developing a national strategy that would define critical products essential for the function of the healthcare system, including emergency response. For these critical products, a supply chain analysis should be performed, and if necessary, contingency plans such as alternative sourcing, inventory bumps, and for the most critical products, consideration given to providing incentives for renewal of domestic manufacturing facilities to orient them toward key products. Both new and existing facilities could also benefit from adoption of, and where needed, pre-competitive investments in research and development intended to enhance technologies in relatively neglected areas of applied pharmaceuticals and manufacturing such as sterile filling and production of injectibles. Models for this approach, including close partnerships with the private sector, have been successful in enhancing pandemic and bioterrorism preparedness, and are also used by the U.S. Department of Defense for their most critical supply chain needs.

A strategy to identify critical needs must also include assessments as to how those needs will change. One of the results of supply chain vulnerabilities has been the problem, particularly in Africa and Asia, of counterfeit and substandard antibiotics, antimalarials, and anti-TB drugs which are very likely not only to harm people directly, but, where sub-potent, help select for resistant organisms. Even where high quality and safe antibiotics are used, we now live in a world where organisms are becoming increasingly resistant to historically effective treatments through inherent genetic selection and gene sharing among pathogens, driven in part by profligate use and misuse of antibiotics in medicine and agriculture, and failures in infection control practices. In 2000, a U.S. task force labeled the antibiotic resistance problem critical, including identifying a market failure requiring consideration of incentives and, fortunately, the issue of AMR is once again at the forefront as indicated by recent White House, WHO and CDC initiatives. While we are beginning to see encouraging, attention to and changes in antibiotic stewardship in the US and other countries in both agriculture and human medicine, in some cases there are now as few as one and sometimes no available drugs to treat resistant organisms. Ominously, there is a largely dry antibiotic drug pipeline driven by the failure to recognize the seriousness of the problem early enough as well as the difficulty of clinical trials (e.g., patient numbers, diagnosis, early empiric therapy) combined with past regulatory uncertainty.

While there are encouraging signs that venture capital firms and pharmaceutical companies are recognizing resistance as a market need and opportunity, the gap remains large and the current growth of antibiotic resistant organisms requires additional government and private sector investment in basic and clinical research in both antibiotics and other approaches. These pathways to new antibiotic development are not mutually exclusive. Government incentives for the industry may be tied to willingness to pay well (as seen for oncology drugs) for important advances, but should also be linked to new approaches promoting phased and appropriate use because broad use will accelerate resistance and shorten the useful life of the drug. Public-private partnering such as recent BARDA[23] investments for antibiotics for drug resistant biothreats, which focus on dual use that can help enhance both national preparedness and medical care, build on successful approaches for preparedness, and should be extended from terrorism and pandemic preparedness to naturally occurring public health threats such as AMR, where there are market gaps.

Public-private engagement may include not only governmental investment but targeted regulatory pathways and market incentives. The Generating Antibiotic Incentives Now (GAIN) Act[24] provides new incentives for needed antibiotics, including five additional years of market exclusivity for qualifying products as well as eligibility for fast-track and priority FDA review. The proposed Antibiotic Development to Advance Patient Treatment (ADAPT) Act[25] would provide explicit authorization for a limited population antibiotic development (LPAD) pathway[26] for smaller specific population/trials and labeling. Even without additional antibiotic-specific legislation, the FDA has existing authority to facilitate review and approval of new antibiotics and is already taking important steps forward in encouraging targeted antimicrobial development (e.g., indications focused on resistant organisms rather than broad clinical entities) and in speeding such development, where initial evidence indicates clinical benefit, like its recent designation and development pathway for new breakthrough therapies (see discussion at http://www.fda.gov/downloads/drugs/guidancecomplianceregulatoryinformation/guidances/ucm358301.pdf).

BIG (MORE AND MORE) DATA: SWIM OR DROWN?

Implicit within many of the foregoing discussions are the capability and role that so-called big data will increasingly play in medical product development and regulatory decision-making. There are tremendous potential opportunities, such as in better understanding diseases, disease and patient subgroups, designing new drugs and vaccines, discovering response biomarkers, tracking the supply chain, and examining product and other medical and public health interventions' safety and performance, the reality is that firms, regulators, and patients alike also increasingly risk drowning in data.

An example is the use of big data by national healthcare agencies, like the U.S. Centers for Medicare and Medicaid Services. CMS collects and reports

23. Jarrett, E., 2014. How is BARDA Positioned to Respond to the Threat? ASPR Blog. Retrieved from http://www.phe.gov/ASPRBlog/Lists/Posts/Post.aspx?ID=96.

24. H.R. 2182, 112th Congress, 2011. Generating Antibiotic Incentives Now (GAIN) Act. Retrieved from https://beta.congress.gov/bill/112th-congress/house-bill/2182.

25. H.R. 3742, 113th Congress, 2013. Antibiotic Development to Advance Treatment (ADAPT) Act. Retrieved from https://beta.congress.gov/bill/113th-congress/house-bill/3742.

26. Karst, K., 2013. ADAPT Act Would Create a New Limited Population Approval Pathway for Antibiotics and Antifungals. FDA Law Blog. Retrieved from http://www.fdalawblog.net/fda_law_blog_hyman_phelps/2013/12/adapt-act-would-create-a-new-limited-population-approval-pathway-for-antibiotics-and-antifungals.html.

quality measures and links these measures with value-based incentive payments for providers.

Yet the sea of data increasingly presents many challenges and problems. In science and in medicine particularly, researchers and policymakers have become both enamored and overwhelmed by the data and its complexity. Data itself does not equate with knowledge and understanding. Advanced computers and computational facilities are now so complex that no one person can build, fix, or understand them; conversely, the actual world of practice has become extraordinarily specialized with some physicians who may care for only one organ. Multiple analyses will and do routinely yield results and associations of $p < 0.05$ that are actually not true and make no sense, in health or science, which can have an opportunity cost of distracting from true issues, findings, and intelligence gathering. Conversely, we are confronted with continuing failures to predict, see, or recognize the real findings amidst the data storm.

Human beings, with interdisciplinary training and working as part of interdisciplinary teams, are going to be increasingly critical for ensuring the effective use of big data. Planners in the government, the academy, and industry must establish and provide resources to support career trajectories for not only specialized experts, but people with feet in more than one world, such as informatics and medicine, and multiple cultures, like medicine and anthropology, as well as generalists with skills to evaluate aggregate phenomena from more than one perspective. Well-informed, generalist participants and leaders of and participants in team big data projects are critically needed to help balance the emphasis on tool- and data-driven science versus science-driven tools and data, while rewarding and prioritizing areas where human contact and analysis—especially the ability to recognize and ask the right questions—and to recognize wrong answers, even when statistically significant, must play essential roles.

CONCLUSION

This chapter has analyzed some of the important trends under way relevant to facilitating the development and access to new medicines, vulnerabilities arising from the expansion of pharmaceutical supply chains worldwide, and some of the new tools scientists, medical providers, regulators, and patients will have at their disposal to address these new challenges. As an overarching matter, many of these trends occur within systems that are viewed by key stakeholders and decision makers as self-contained and discrete. They are not. Harnessing new technology and data analytics to both develop new drugs and fight antimicrobial resistance will require broad, systematic approaches to dealing with highly complex issues. That starts with beneficial collaboration and discussions including leaders and experts in business, law, foreign affairs, medicine, health systems, U.S. and global regulation and regulatory science, and global public health. These collaborations must welcome diverse opinions, and multisector engagement in a manner where differences and new ideas can be freely aired and debated. Doing so will enhance understanding across sectors and silos and facilitate unique and win-win solutions that improve global health.

Chapter 2

FDA's Global Investigation and Enforcement Authority, Partnerships, and Priorities

Marc J. Scheineson
Food and Drug Law Practice Group, Alston & Bird, LLP, Washington, D.C., USA

Chapter Outline

INTRODUCTION

FDA Commissioner Margaret Hamburg, M.D., dedicated her tenure, the second longest of any modern FDA commissioner, to improving product quality. This mantra became one of the top priorities of an agency regulating drugs, biologics, medical devices and over 25% of the American economy. She soon realized that 40% of all finished drugs used in the United States, and 80% of active ingredients, were made outside the United States. Similar statistics applied to other FDA-regulated products including processed food, medical and radiological devices, cosmetics, and consumer personal care products. This realization unleashed an unprecedented parade of FDA inspectors throughout U.S. manufacturing establishments and to the farthest corners of China, India, South America, and the European Union. The FDA reported in budget documents that in fiscal year (FY) 2015, overseas inspections are expected to rise from 827 inspections to 999, while domestic facility inspections rise from 1851 to 1856.[1] This regulatory action follows years of increasing inspections that included 808 high-risk facilities in FY2013, including 443 domestic drug inspections,

1. Department of Health and Human Services, Food and Drug Administration, 2015. Justification of Estimates for Appropriations Committees. Available at http://www.fda.gov/downloads/AboutFDA/ReportsManualsForms/Reports/BudgetReports/UCM388309.pdf.

Food and Drug Regulation in an Era of Globalized Markets. http://dx.doi.org/10.1016/B978-0-12-802311-2.00002-3

and 365 foreign drug establishments. The scope and pace of manufacturing audits required the FDA to, at last, team with foreign regulators including the European Medicines Agency (EMA) on generic drug site inspections, and with local inspectors throughout the world.

The FDA has always been considered the "gold standard" of product application review and manufacturing standards, at least by public health officials in the United States. This chapter contains a summary of these standards and regulatory requirements as they relate to products marketed and sold in the United States, even if those products are actually made outside of the United States. This summary focuses primarily on the manufacturer of prescription drugs because those products are the most heavily regulated. The same or similar regulatory structures also apply to other healthcare products regulated by the FDA under the authority of the Federal Food, Drug, and Cosmetic Act (FDCA)[2] including processed foods, other than beef, poultry, and eggs (regulated by the U.S. Department of Agriculture); medical and radiological devices; biologics; over-the-counter drugs; animal food and drugs; cosmetics; and consumer care products.

ADULTERATION, MISBRANDING, AND GMPS

The structure of the U.S. FDCA that established the FDA and defines its authority, is based on determining if a product is "adulterated" or "misbranded." A drug is deemed adulterated if (1) it contains any filthy or decomposed substance[3]; (2) is prepared, packed, or held under unsanitary conditions[4]; or (3) methods used to make the product do not conform to current good manufacturing practices (cGMPs).[5]

The adulteration statutory provision administered by the FDA provides further that a product is adulterated if (1) its container is composed of poisonous or deleterious substances that may cause the contents to be injurious to health[6]; (2) the product contains an unsafe coloring additive[7]; (3) the strength, quality, or purity falls below compendial standards[8] (United States Pharmacopeia or other similar publications); (4) the strength or purity falls below what it purports to possess[9] (in product labeling, specifications or otherwise); or (5) it is mixed or packaged to reduce its quality or strength.[10]

---

2. 21 U.S.C. § 301 *et seq.*
3. § 501(a)(1), Food, Drug, and Cosmetics Act; 21 U.S.C. § 351(a)(1).
4. *Id.* at § 501(a)(2)(A); § 351(a)(2)(A).
5. *Id.* at § 501(a)(2)(B); § 351(a)(2)(B).
6. *Id.* at § 501(a)(3); § 351(a)(3).
7. *Id.* at § 501(a)(4); § 351(a)(4).
8. *Id.* at § 501(b); § 351(b).
9. *Id.* at § 501(c); § 351(c).
10. *Id.* at § 501(d); § 351(d).

A drug is deemed "misbranded" if (1) its labeling is false or misleading (in any particular)[11]; (2) the label fails to bear required information including the name and place of business, accurate statement of contents, weight, measure, or numerical count[12]; (3) required labeling is not prominently displayed or understandable[13]; (4) it does not contain the established name and names and quantities of active ingredients[14]; (5) labeling does not bear adequate directions of use[15]; (6) the drug is not packaged or labeled as prescribed in the official comprendium[16]; (7) reference is not made to registration or registration numbers denoting FDA approval of the firm or products[17]; (8) the drug is manufactured in a nonregistered establishment[18]; or (9) it is not included in the drug listing.[19]

The concept of GMPs that are included in the FDCA rely on the following principles:

(1) Quality, safety, and effectiveness must be designed into a product;
(2) Quality cannot be inspected or tested into a finished product; and
(3) Each step of the manufacturing process must be controlled (and documented) to maximize the likelihood that the finished product is acceptable.

A drug manufacturer is required to establish and maintain methods of GMPs, and facilities and controls used for the manufacture, processing, packing, or holding of a drug to assure that the drug is (1) safe; (2) has appropriate identity and strength; and (3) meets quality and purity characteristics.[20] FDA regulations defining GMPs are general. They are tailored to individual manufacturing establishments. The FDA GMP regulations are divided into specific categories including:

1. *Buildings and Facilities*. Product manufacturing must be confined to separate or defined areas of adequate size necessary to prevent contamination or mix-ups. Air filtration systems (HVAC) are required to be installed in all production areas. Stringent sanitation standards apply.[21]
2. *Components, Containers, and Closures*. Written procedures must exist for the receipt and testing of all incoming materials. These manufacturer standard operating procedures (SOPs) must include active ingredients;

11. § 502(a), Food, Drug, and Cosmetics Act; 21 U.S.C. § 352(a).
12. *Id.* at § 502(b); § 352(b).
13. *Id.* at § 502(c); § 352(b).
14. *Id.* at § 502(e); § 352(e).
15. *Id.* at § 502(e)(1)(B); § 352(e)(1)(B).
16. *Id.* at § 502(f); 352(f).
17. *Id.* at § 502(g); 352(g).
18. 21 C.F.R. §207.39.
19. § 502(o), Food, Drug, and Cosmetics Act; 21 U.S.C. § 352(o); § 510(j), Food, Drug, and Cosmetics Act.
20. 21 C.F.R. Parts 210, 211.
21. *See* 21 C.F.R. §§ 211.42-211.58.

excipients (coating, binder, flavor preservative, etc.); and containers and closures.[22] Procedures must also exist for receipt and storage of manufacturing materials,[23] and their testing, approval, and rejection.[24]

3. *Production and Process Controls*. The manufacturer must maintain written production and process control procedures. These SOPs must be approved by a quality unit. They must be followed closely in production and must be documented at the time of performance. Any deviation from these written procedures must be recorded and justified in the written production logs.[25]

Maintenance of cGMPs is largely at the discretion of the drug manufacturer and FDA inspectors, who frequently hold diverging views as to the nature and extent of manufacturers' legal obligations. In a lengthy dispute between a global specialty pharmaceutical company based in the United States and the FDA over the precise content and requirements of good manufacturing practices, a U.S. federal trial court provided a comprehensive review of cGMP procedures in the landmark case of *U.S. v Barr Labs, Inc.*[26] The issues in that case related to the sufficiency of a generic drug-makers compliance with cGMPs and the extent of the FDA's authority in the field without being deemed arbitrary, capricious, or otherwise abusing its discretion. The court took the unusual course of delving into the technical details of pharmaceutical manufacturing processes and practices. The result was a primer on the procedural requirements of each component of the FDA's cGMP regulations and the FDA's authority to request a recall to the extent that GMP deficiencies are observed. Barr Labs was ultimately ordered by the court to produce data underlying its analytical validation studies and agree to withhold further production until it could complete a process of prospective validation of manufacturing processes satisfactory to the FDA.

Following the *Barr Labs* order, the FDA issued a series of guidance documents related to use of out-of-specification product, withdrew proposed cGMP regulations (and then issued amendments to its existing cGMP regulations for finished pharmaceuticals),[27] and has adopted a twenty-first century risk-based model for prioritizing sites for manufacturing inspections.

FDA ESTABLISHMENT INSPECTIONS

1. Statutory Authority to Conduct Establishment Inspections

The FDCA provides the FDA with express authority to inspect all registered manufacturing establishments subject to some general limitations.[28] For example,

---

22. *Id*. at § 211.80.
23. § 211.82.
24. § 211.84.
25. § 211.100.
26. 812 F.Supp. 458 (D.N.J. 1993).
27. 73 F.R. 51,919 (effective Dec. 8, 2008).
28. § 704, Food, Drug, and Cosmetics Act; 21 U.S.C. §§ 374, 360(h).

inspections must be conducted at reasonable times, within reasonable limits, and in a reasonable manner. Drug establishments must be inspected every 2 years.[29] Manufacturer consent is unnecessary, as are search warrants, *Miranda* warnings, and inspection warrants.

The FDA possesses no general subpoena authority, but in certain civil proceedings, it may issue subpoenas for the production of evidence and records.

2. Scope of Inspections

FDA was expressly given the authority to inspect and copy the following: records, files, papers, processes, controls, and documents related to a facility's bearing on whether drugs are adulterated, misbranded, or otherwise violative.[30] At the same time, FDA authority does not allow its investigators to inspect an establishment's (a) financial data; (b) sales data other than shipping figures; (c) pricing data; (d) personnel data other than information establishing qualifications of technical and professional personnel; (e) and certain research data except to the extent that such information is required for inspection or is submitted to the FDA for particular products.[31]

The FDA is not expressly authorized to take photographs, but often attempts to persuade facilities being inspected to allow field investigators to make photographs within facilities being inspected. The agency often cites two judicial decisions: *Dow Chemical Co. v. United States*, 476 U.S. 227 (1986) (holding that the Environmental Protection Agency had statutory authority to use aerial photography during a site inspection of a chemical company's industrial complex) and *United States v. Acri Wholesale Grocery Co.*, 409 F. Supp. 529 (S.D. Iowa, 1976) (holding that photographing warehouse conditions by FDA agents was not unreasonable). Neither case, however, directly addresses the scope of FDA authority to take photographs during an in-plant inspection. The *Dow Chemical* Case involved aerial photography taken of the outer facade of the facility only and not the inside of a manufacturing plant that may generally be deemed as proprietary. The *Acri Wholesale Grocery* ruling focused on the company's failure to object to FDA inspectors' requests for photography in a timely manner and not on whether the FDA was empowered to take photographs. Despite the absence of case law directly supporting the agency's right to take photographs, its updated *Investigations Operations Manual* instructs FDA inspectors not to request permission to take photographs and encourages them instead to assert their right to take photographs.[32] In recent nonbinding draft guidance, the FDA also states that any attempt to limit photography during inspection will constitute an unacceptable obstruction resulting in detention of products and potential

---

29. 21 U.S.C. § 360(h).
30. § 704(a)(1), Food, Drug, and Cosmetics Act; 21 U.S.C. § 374(a)(1).
31. *Id.*
32. *See* Food and Drug Administration. FDA Inspections Operations Manual, §5.3.4.1; Cited in Tsekerides, T., 2013. Product Liability Monitor. Available at http://product-liability.weil.com.

prevention of sale of any drugs seized by the FDA.[33] The FDA's proposed examples of prohibited behavior often hinge on whether "reasonable explanations" or "adequate justifications" exist to explain the limitations imposed on inspectors attempting to photograph the inspection site.[34]

FDA field investigators also routinely seek to have responsible officials' written affidavits. Those legal documents might stipulate that the company ships goods in interstate commerce (to establish FDA jurisdiction), or may admit to particular violations of the FDCA or other prohibited acts. There is no authority under §704, or other sections of the FDCA, authorizing the FDA to obtain such affidavits. As these written admissions are legal documents that could adversely affect judicial proceedings if a matter was litigated, clients are advised not to sign anything presented during FDA field inspections without the review of regulatory counsel. This policy should be documented in written SOPs maintained by FDA-regulated establishments relating to the process used in the event of FDA inspections. FDA field investigators will quickly back off any pressure to sign such documents when presented with such written policy.

The FDCA does not expressly authorize FDA investigators to interview company employees with or without company counsel present. While it is advisable to cooperate and maintain cordial professional relations with FDA investigators who wield significant enforcement discretion, a company should maintain responsibility and control of the inspection process. For this reason, many companies limit FDA interviews to selected managerial employees only, and may include regulatory counsel.

3. Procedure for Conducting Inspections

The procedures used in establishment inspections are generally dictated by the FDA *Investigation Operations Manual*.[35] This publication, available on the FDA website, instructs FDA field investigators concerning agency policy with respect to specific factual elements of the inspection. Knowing how the field investigator thinks provides important clues to how to prepare and help manage the facility inspection.

Those procedures generally commence with a team of FDA field investigators appearing at the door of an establishment registered to make FDA-regulated products. Their appearance may be either unannounced in the case

---

33. *See* Department of Health and Human Services, Food and Drug Administration, Office of Regulatory Affairs, Center for Drug Evaluation and Research, Center for Biologics Evaluation and Research, Center for Veterinary Medicine, 2013. Guidance for Industry Circumstances that Constitute Delaying, Denying, Limiting, or Refusing a Drug Inspection. Available at http://assets.law360news.com/0456000/456930/ORA20138%20FDASIA%20707%20Guidance%20FINAL-%20July%2012%202013-508.pdf.

34. Overley, J., 2013. FDA Warns Against Delaying Inspectors, Details Drug Seizures. Law 360. Available at http://www.law360.com/articles/456930/print?section=lifesciences.

35. *See* Food and Drug Administration. Investigations Operations Manual. Available at http://www.fda.gov/iceci/inspections/iom/default.htm.

of drug manufacturing facilities, or announced and coordinated with company managers in advance in the case of medical device or foreign product manufacturers. FDA investigators should appear only during normal business hours, or may be denied admittance. The lead investigator will present credentials (e.g., big, gold, official-looking FDA badge) and a signed FDA Form 492 "Notice of Inspection." The most common reasons that the FDA inspects a manufacturing facility include (a) a routine inspection required every 2 years by statute; (b) for cause; (c) follow-up of a warning letter or enforcement action; (d) complaints from consumers, industry competitors, or current or former employees (e.g., whistleblowers); (e) effectiveness checks following a voluntary recall to determine the inclusiveness of the recall; (f) preapproval inspections required before Rx drug approvals; or (g) special enforcement initiatives (e.g., compounding pharmacies, generic drug makers, etc.).

Following the inspection, which can take a day or 2 or many weeks depending on its scope, the inspector will conduct an exit interview with the most senior responsible party on the premises. The company will likely be presented with a Form 483 listing in detail the deficiencies noted during the inspection. The company routinely has 15 calendar days to respond to the Form 483, noting the steps it will take to address those observations. Following completion of the inspection and resolution of noted deficiencies, the investigation will be closed by the FDA. At that time, the FDA will send the company a formal Establishment Inspection Report (EIR), if requested. This document contains valuable information concerning the process used to conduct the inspection and details concerning the FDA's view of the company's quality systems (as required by FDA cGMP regulations).

A registered establishment may not refuse to admit an FDA field investigator without cause. Such a refusal could be treated, at the FDA's discretion, as a "prohibited act" under §301 of the FDCA. Such action could result in the issuance of an injunction against the regulated company, civil or criminal penalties, or issuance of a judicially enforced administrative warrant allowing the inspection. Refusal in the face of a warrant could constitute "contempt of court" and lead to judicial proceedings and imprisonment.

At the end of the inspection, the lead investigator (if more than one visited the premises), will conduct an exit interview with the most senior responsible official for the company. During this wrap-up discussion, the manager will often be presented with a written Form 483 containing a list of noted "observations" (deficiencies). During that interview, it is often possible for the company manager to make sure the observations are valid and accurate. For example, if documents are cited as unavailable or incomplete, the manager can provide those records if they were overlooked or misinterpreted. Some proposed observations may also be eliminated if, for example, the inspector can be persuaded that the underlying facts are incorrect, or corrective actions have already been taken. Disagreements with observations can be noted by the investigator. The company can also obtain clarification concerning the findings to assist in preparing an adequate response and eliminating misunderstandings. This also provides an opportunity to propose corrective actions, as needed, and gauge whether they are likely sufficient in the eyes of the investigator. While Form 483s are not commonly posted publicly on the FDA website, as are warning letters, they can be requested by competitors, or others, using FDA's Freedom of Information Act procedures under 21 C.F.R. §20.20.

When investigators return to the FDA District Office, they begin preparing a much more detailed EIR. That document assists to inform their supervisors, and often the relevant product center leadership at FDA headquarters in White Oak, MD (suburb of Washington, DC), concerning the details and severity of the inspection results. The company's response to the Form 483 is expected within 15 calendar days of its issuance. It is expected that the response will contain detailed explanations concerning how observations and quality systems will be corrected, including specific timelines and support documents. If the company response to the 483 is determined to be insufficient, or the underlying observations are deemed to be sufficiently serious, the FDA may issue a Warning Letter based on the input of agency compliance officials. At times, these Warning Letters are issued even before the company's response can be delivered or reviewed.

FDA ENFORCEMENT INSTRUMENTS RELATED TO MANUFACTURING

The FDA is granted the authority to use a variety of mechanisms to insure enforcement of its cGMP manufacturing rules. Other than Form 483 Notice of Deficiencies discussed above, other instruments to assure compliance include (1) Warning Letters citing continued deficiencies that are posted publicly on the FDA website and require company response within 15 calendar days; (2) the adverse publicity that can come from publicly posted documents including, on occasion, Safety Alerts; (3) use of product recalls that may either be voluntary by the manufacturer or requested by the FDA[36]; (4) suspension or withdrawal of product approval; (5) adverse impacts on the awards of government contracts, including purchase of impacted products by federal health insurance programs; (6) seizures; (7) injunctions and consent decrees; and (8) in limited cases, federal criminal prosecutions.

FDA statistics indicate that the issuance of Warning Letters related to product quality issues (for medical devices alone) increased dramatically under the administration of Commissioner Hamburg from approximately 93 in 2010 to 169 in 2012 (and many more in 2013-2014, but final statistics have not been compiled).[37] The bulk of the cGMP deficiencies cited by the FDA

36. *See* 21 C.F.R. §7.45.

37. March 2013. FDA Reports. The Silver Sheet.

in these Warning Letters for medical devices, for example, have included 111 (66%) related to improper investigation of complaints (66%); 104 (62%) related to inadequate corrective action and preventative action plans (CAPA); 90 (53%) design controls; 81 (48%) failure to file adequate Medical Device Reports (for injuries caused by medical devices); 70 (41%) acceptance activities; 60 (35%) related to purchasing controls; 51 (30%) production and process controls; 51 (30%) device history records; 50 (29%) process validation; and 43 (25%) quality audits.

The FDA has only maintained separate statistics for CDER drug Warning Letters by the issuing office since 2011. Since that time, Warning Letters related to drug manufacturing and product quality issued by its Office of Manufacturing and Product Quality have been consistent with 19 issued in 2014; 21 issued in 2013; 22 issued in 2012; and 20 issued in 2011.

FOREIGN INSPECTIONS

In order for a foreign manufacturer to market and ship FDA-regulated products to the United States, it must register its manufacturing establishment on the FDA Unified Registration and Listing System (FURLS) database. FDA-regulated products shipped to the United States that do not contain verified current establishment registration numbers are denied entry into the United States at their ports of arrival. Once a manufacturing establishment is registered on FDA's electronic FURLS database, it is subject to inspection either by FDA field investigators or, in limited cases, by foreign agency investigators pursuant to memoranda of understanding with the FDA.

Beginning in FY 2013 when the FDA Safety and Innovation Act (FDASIA) became effective, the FDA was required to prepare annual reports concerning registered establishment inspections for both domestic and foreign locations.[38] FDA's inspection force was expanded and detailed to U.S. Embassies around the world. In FY2013, 9120 domestic drug establishments were registered, and 12,468 domestic device establishments. At the same time, 3493 foreign drug establishments were registered and 9385 foreign device establishments. Of these registered manufacturing establishments, 4858 domestic inspections were conducted and 1138 foreign inspections were conducted.

CONCLUSION

The system for the manufacture of FDA-regulated products, including drugs and medical devices, is clearly global. Nearly 40% of all finished drugs used in the United States, and 80% of active ingredients, are now made outside the United States, but are within the FDA's jurisdiction if those drugs are marketed

38. § 510(h)(6), Food, Drug, and Cosmetics Act.

and sold in the United States. Similar statistics apply to other FDA-regulated products. The FDA is conducting a large number of inspections of domestic and foreign manufacturing establishments as the top agency priority, based on revised electronic registration laws and a deployment of FDA inspectors aboard. For this reason, it is more important than ever for domestic and foreign manufacturers of FDA-regulated products to learn the rules and become compliant with FDA cGMP regulations.

Chapter 3

The European Medicines Agency and the Regulation of Medicines in the European Union

Sabine Haubenreisser*, Martin Harvey-Allchurch*
*International Affairs, European Medicines Agency, London, United Kingdom

Chapter Outline

INTRODUCTION

Long before the era of globalized markets that we now take for granted, Europe had put together one of the most durable and effective interstate drug regulatory processes currently in operation. Dating back to January 1965, the European medicines regulatory system now brings together 28 member states of the European Union (EU) in addition to Iceland, Liechtenstein, and Norway.

This remarkable example of intraregional cooperation not only creates a single market of some 500 million patients across 31 countries, but more importantly is a powerful vehicle for pursuing public and animal health objectives.

APPROVAL OF DRUGS IN EUROPE

In the EU, a drug can only be placed on the market after a marketing authorization has been granted. This premarketing approval can either come from the

Food and Drug Regulation in an Era of Globalized Markets. http://dx.doi.org/10.1016/B978-0-12-802311-2.00003-5

competent authority of an EU member state or from the European Commission (EC) based on the opinion of the European Medicines Agency (EMA).

Today, there are four different routes for marketing authorization: three national routes and one European centralized route. The choice of route can depend on a number of factors: the type of product, the authorization history in the EU, as well as regulatory and marketing strategy or even company preferences.

Each EU member state has its own national procedures for the authorization within their own territory of medicines that fall outside the scope of the centralized procedure. In addition there are two possible routes available to sponsors for the authorization of these medicines in more than one country. The "decentralized" procedure allows companies to apply for the simultaneous authorization of a medicine in more than one EU country if it has not yet been authorized in any EU country and it does not fall within the mandatory scope of the centralized procedure. Under the "mutual recognition procedure" companies that have a medicine already authorized in one EU member state can apply for this authorization to be recognized in other EU countries. If there are disagreements that cannot be resolved between member states, disputes arising in these procedures can be referred to the EMA for arbitration that leads to a Europe-wide decision.

Almost all new innovative medicines today are, however, authorized through the EU-wide centralized procedure. Under the centralized procedure, there is one single application, one single assessment, and one single marketing approval for the whole of the EU, and which also extends to Iceland, Lichtenstein, and Norway.

In addition to single approval for the whole of the EU, products authorized through the centralized procedure are approved with one trade name and one label. This includes the same summary of product characteristics for healthcare professionals and package insert for patients, translated into all 24 official EU languages.

The centralized procedure is primarily aimed at new innovative medicines, most of which are legally obliged to use the procedure either because it is a "biotech" product (e.g., recombinant DNA technology), because of the therapeutic class (e.g., AIDS/HIV, cancer, diabetes, neurodegenerative disorders), or because of the type of product (e.g., orphan drug, advanced therapy drug). Other products may use the centralized procedure where the applicant can demonstrate significant therapeutic, scientific, or technical innovation, or show it is in patient or animal health interest at the EU level.

Once the EMA adopts its opinion on whether or not a drug should be approved, the legal decision to grant the EU-wide marketing authorization is taken by the EC. The EC also adopts the legal decision where disputes between member states in the mutual recognition or decentralized procedure are referred to the EMA.

THE EUROPEAN MEDICINES AGENCY: ITS ROLE AND ACTIVITIES

The EMA began life in January 1995. Originally known as the European Agency for the Evaluation of Medicinal Product, its name was simplified to EMA in 2004 to reflect its growing mandate. It is based in London, UK.

The EMA is an independent agency of the EU. This means that it is not part of the EC or any other EU institution, although it works in very close cooperation with the EC. Rather than creating a central "federal" body, the EMA is structured as a networking body that provides an interface of cooperation and coordination of member states' activities with respect to medicines. It has a scientific secretariat of about 800 highly trained experienced scientists, lawyers, and other staff, and one of its main functions is to coordinate and bring together the expertise and resources put at the agency's disposal by the member states rather than replicate them at a central level.

An overview of the EMA's main activities and responsibilities is given below, together with an overview of some of the areas in which it does not have a mandate.

The EMA's main activities include:

- Management of the centralized procedure
- Human and veterinary, chemical, and biological products
- Evaluation of new products, generics, OTCs
- Assistance on scientific advice and protocol assistance to sponsors
- Orphan drug designation
- Pediatric investigation plans (PIPs)
- Regulatory and scientific guidance
- Advanced therapy medicines
- Herbal medicines monographs and a list of herbal substances
- Information to patients and transparency
- Arbitration and referral procedures for nationally approved medicines
- Coordination of EU pharmacovigilance activities
- Coordination of member states' inspections (GMP, GCP, GLP)
- Advice to European Commission and member states on any issue relating to drugs

The EMA is not responsible for:

- Research and development
- Analytical testing of samples, inspections
- Food
- Tobacco
- Clinical trials authorization
- Ethics committee
- Patents and intellectual property rights
- Pricing and reimbursement
- Advertisement and promotion
- Medical devices (except combination products; e.g., drug with delivery device, device containing an ancillary active substance)

The EMA finds itself in the center of a European network of some 45 national, competent authorities coming from 28 EU member states, Iceland, Liechtenstein, and Norway, and the EC. This network has a pool of over 3500 European experts drawn from across the EU coming from national authorities, hospitals, academia, research centers, and patient representatives.

It is the contribution of these European experts that underpins the work of the EMA's committees and working parties.

The EMA attaches a great deal of importance to ensuring that it is able to provide the best possible scientific advice to the EC and member states, as well as for sponsors. The scientific competence of each European expert is guaranteed by their nominating authority, and their independence and integrity is assured through a public declaration of interests. European experts do their work for the EMA independently of their nominating authority. Their declarations of interest forms are published on the EMA's website and must be updated each year, or more frequently as their interests change. Every meeting begins with a review of all participants' conflicts of interests.

The same robust conflict of interest regime applies to EMA staff members, who are also required to make public declarations of interest.

The EMA seeks to operate as a "one-stop shop" for sponsors who wish to bring a drug to the EU market. It offers joined-up support opportunities right through the product research and development cycle; from scientific advice and protocol assistance, orphan drug designation, and pediatric development plans, through to postauthorization safety and efficacy trials and risk management plans. For small- and medium-size companies, additional support is available to help navigate the regulatory process.

The EMA operates through seven scientific committees and a number of working parties. These are largely made up of experts nominated by member states, but four of the seven committees also have members explicitly nominated to represent patients and healthcare professionals. All members play an equal role in the work of the committee or working group. Their work can be supplemented in some instances by scientific advisory groups to advise on questions relating to the evaluation of specific types of medicines or treatments. These groups consist of European experts selected and nominated according to the particular expertise required.

EMA Scientific Committees:

- Committee for Medicinal Products for Human Use (CHMP)
- Pharmacovigilance Risk Assessment Committee (PRAC)
- Committee for Medicinal Products for Veterinary Use (CVMP)
- Committee for Orphan Medicinal Products (COMP)
- Committee on Herbal Medicinal Products (HMPC)
- Committee for Advanced Therapies (CAT)
- Pediatric Committee (PDCO)

Current CHMP standing and temporary working parties and expert groups:

- Patients' and Consumers' Working Party
- Healthcare Professionals' Working Party
- Scientific Advice Working Party
- Biologics Working Party
- Quality Working Party
- Safety Working Party

- Biosimilar Medicinal Products Working Party
- Biostatistics Working Party
- Blood Products Working Party
- Cardiovascular Working Party
- Central Nervous System Working Party
- Infectious Diseases Working Party
- Oncology Working Party
- Pharmacogenomics Working Party
- Pharmacokinetics Working Party
- Rheumatology/Immunology Working Party
- Vaccines Working Party
- Additional Scientific Advisory Groups

EMA SUPPORT IN THE EARLY STAGES OF DRUG DEVELOPMENT

As new medicines become more complex to assess and the financial implications to the pharmaceutical industry spiral up, not to mention the opportunity costs for public health, the EMA has sought to intervene earlier in the drug discovery cycle (Figure 3.1).

This has the advantage of allowing the regulator to learn more about the product at an early stage and to offer advice, and for sponsors to know in advance any issues in the development plans that may cause concern for regulators. It can also

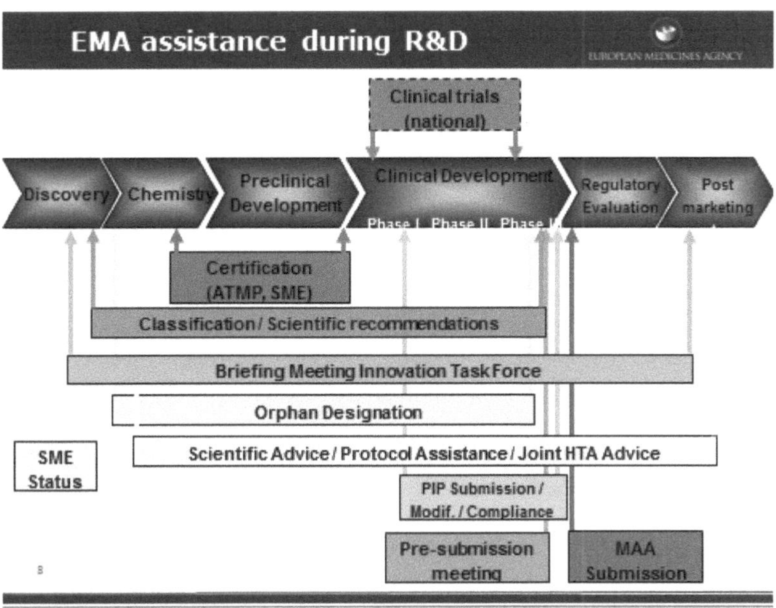

FIGURE 3.1 EMA assistance during R&D.

avoid inappropriate exposure of patients in clinical trials that may have incorrect design or may even not be needed at all.

The Agency also offers significant resources and guidance to sponsors, whether micro, small, medium or large entities, to help them prepare and submit applications. Much of this is available on the EMA website (www.ema.europa.eu) in addition to the possibility of face-to-face meetings offered through the Agency's various support programs. There is also a legal requirement of all applicants to have agreed to a pediatric development plan with the EMA during the drug development phase, usually around the end of Phase 2 studies. The EMA cannot accept an application for review without compliance with an agreed upon PIP, or if a waiver or deferral has been agreed to.

Early identification of orphan drug candidates allows formal designation, which in turn gives access to special programs for protocol advice and other incentives, both at the European and the national level.

This early interaction allows the EMA and the sponsor to map out a pathway through the development, presubmission, and approval phases. This has proven to be especially successful to allow the EMA to support innovative small- and medium-sized enterprises and academic spin-offs, but also for highly innovative advanced therapy products. It also allows for discussion of the possible use of special approval pathways such as conditional or accelerated approvals.

A recent analysis of the impact of scientific advice on the outcome of marketing authorization applications between 2008 and 2012,* suggests strongly that compliance by sponsors with advice – especially early advice – on clinical trial design had a positive correlation with a positive outcome at approval and also time to approval.

THE CENTRALIZED PROCEDURE AT WORK

The European centralized procedure is characterized by its single application, single evaluation, single authorization model (Figure 3.2).

Assessment Overview

The single evaluation review is carried out by the Committee for Medicinal Products for Human Use (CHMP). The CHMP has one representative from each of the EU member states, Iceland, Liechtenstein, and Norway, together with up to five additional members co-opted to complement the scientific expertise of other nominated members; for example, the current co-opted members have specialisms in biotechnology, statistics, pharmacovigilance and epidemiology. The CHMP has a legal time limit of 210 days in which to complete the review cycle and deliver its opinion, which can be either positive or negative.

---

* *Regulatory watch: Impact of scientific advice from the European Medicines Agency,* Nature Reviews Drug Discovery (2015) doi:10.1038/nrd4621, published online 17 April 2015.

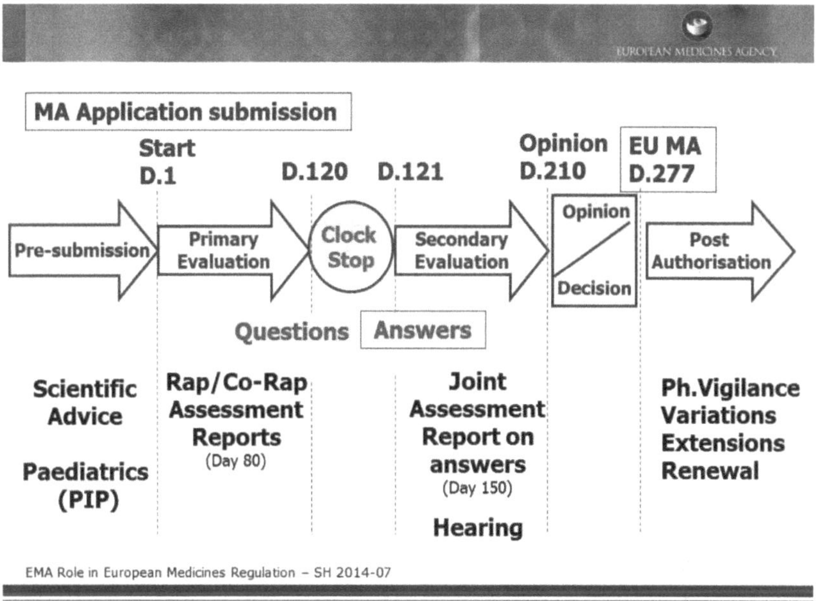

FIGURE 3.2 Overview of centralized procedure.

This is then passed to the European Commission, which takes the final decision whether or not to grant a marketing authorization.

For each application for marketing authorization, two CHMP members are selected to act as rapporteur and co-rapporteur. Each of them works alongside a team of assessors drawn from the list of European experts. For practical purposes, often these assessors will come from the rapporteur's home country, but increasingly multinational expert teams are used, especially where specialist skill sets are required. The rapporteur and co-rapporteur are supported in the procedure by the EMA scientific secretariat.

The rapporteur and co-rapporteur work independently of each other in the assessment of the application. As the lead reviewers on behalf of the CHMP, they coordinate input from external experts and *ad hoc* groups, propose objections and questions, identify the need for inspections, and propose scientifically justifiable labeling for the drug where the original proposed by the applicant cannot be supported.

The role of the rapporteurs is to guide the CHMP in its decision making. While the rapporteur and co-rapporteur's initial assessment reports are highly influential in the process, the final opinion on whether or not to recommend a marketing authorization is a collegial decision adopted by the CHMP as a whole. Decisions are generally taken by consensus, although the Committee does vote if necessary; dissenting opinions are included in the assessment report to ensure full transparency.

Single Assessment Report

The CHMP opinion is based on a single assessment report, which forms an integral part of the formal opinion and reflects the overall scientific evidence that formed the basis for the opinion. It also describes the committee's discussions and any formal hearing of the sponsor.

The report is a public document, intended to explain the reasons for the CHMP's decision. Once the authorization process has been completed, the report is published on the EMA website in the form a European public assessment report (EPAR). For this the EMA may redact, if suitably justified by the applicant, certain elements on the grounds of commercial confidentiality or protection of personal data.

EPARs are systematically published for all drugs reviewed by the EMA regardless of whether the outcome is positive or negative, and are regularly updated throughout the drug's life cycle as more data becomes available, new indications added, warnings included in the label, and so forth.

An EPAR includes a comprehensive summary of the quality, safety, and efficacy of data submitted by the applicant, as well as of the assessment and the CHMP's conclusion. The report is based on the rapporteur and co-rapporteur's own assessment reports and the CHMP discussions, including the committee's evaluation of the balance of benefits and risks. It also reflects the outcome of any inspections requested by the CHMP and consultations with scientific advisory groups or working parties. It is jointly prepared by the EMA scientific secretariat and the rapporteurs.

Patients in the Scientific Review Process

In line with the EMA's strategy to improve the contribution of patients in its work, a pilot project was launched in September 2014 to involve patients in the assessment of the benefits and risks of medicines in the CHMP. Patients are already included in many product review activities, for example scientific advice, orphan drug designation, pediatric development plans, advanced therapies, review of product information prior to approval, and safety communication.

As part of the pilot, patients will be invited to present their views on medicines under review for which there is an unmet medical need and where the CHMP is still undecided. Patients may also be invited to give their views in cases where the committee is considering whether to recommend the maintenance, suspension, or revocation of a marketing authorization, or a restriction of indication of an authorized medicine. The ultimate objective is to explore how patients can be involved systematically and effectively in the work of the CHMP.

TRANSPARENCY AND THE CENTRALIZED PROCEDURE

Transparency is a cornerstone of the EMA's operations and, in particular, to the work of all EMA committees. While all committee and working party meetings are currently conducted entirely in private, the agendas and minutes of all meetings, together with an immediate publication of meeting outcomes and key decisions, are public.

The EMA's transparency policy applies equally to positive and negative outcomes, and to submissions that are withdrawn by applicants before a final decision is taken.

In addition to EPARs, agendas, minutes, broadcasting of workshops, and a number of other initiatives, the EMA also publishes a list of all applications currently under review. The agency adopted a landmark policy in October 2014 that will see it publish the clinical reports that underpin its decision making on medicines. The policy entered into force on January 1, 2015, and applies to clinical reports contained in all applications for centralized marketing authorizations submitted after that date. The reports will be released as soon as a decision on the application has been made.

Another recent innovation is the possibility for the PRAC committee to hold public hearings as part of certain safety reviews of medicines, particularly in relation to their therapeutic effects and therapeutic alternatives available, as well as the feasibility and acceptability of proposed risk management and minimization activities. By opening up the evaluation process and engaging in a dialog with the communities most concerned or affected by certain recommendations, public hearings can improve the public's understanding of the benefit-risk considerations that are a part of regulatory decision making, thereby contributing to increased public confidence in regulatory decisions.

EMA–FDA COOPERATION

The EMA operates in a number of multilateral international forums and has confidentiality arrangements with counterpart authorities around the globe. Of these, the EMA's relationship with the U.S. Food and Drug Administration (FDA) is the most important, both strategically and operationally.

The EMA-FDA agreement was signed in 2003 and extended indefinitely in 2010. The agreement sets out what can be exchanged, especially about ongoing evaluations and safety concerns for medicines. The two agencies can exchange staff and observe the work of each other's scientific committees. The agreement foresees the possibility to second liaison officials between the agencies, where these senior staff members play a key role in ensuring the identification of issues of concern to the two agencies. It facilitates the effective exchange of information and promotes a better understanding of the agencies' operations, fostering convergence.

Cooperation in the Product Life Cycle

The two agencies cooperate right across the product life cycle from drug development, approval, and postmarketing surveillance. Cooperation relating to product development is one of the busiest areas and operates through so-called cluster groups. In the preapproval phase, there are two important clusters dealing with orphan drugs and pediatric drugs, as well as the possibility of providing parallel scientific advice.

The orphan drug cluster is a platform for discussion about common issues of shared interest. It helps to reduce administrative and regulatory burden as well as sharing expertise and exchanging views. One of the most important outcomes of the orphan drug cluster is the agreement on a common submission format for designation as an orphan drug and a common template for the annual reports required by both agencies. There are monthly teleconferences of this cluster.

The pediatric cluster also has representatives from Health Canada, Australia's Therapeutic Goods Administration (TGA), and Japan's Pharmaceutical and Medical Devices Agency (PMDA) and Ministry of Health, Labor and Welfare (MHLW), all of which have confidentiality agreements with the EMA and FDA. This cluster is used to share information on EMA PIPs as well as FDA written requests on waivers. For identified products both EMA and FDA draft common commentaries, which are shared with the sponsor. It seeks to encourage global pediatric development plans that are compatible with the requirements of the agencies.

An important example of this is the 2014 joint EMA-FDA proposal to facilitate clinical investigation of new medicines for Gaucher disease in children. In addition to facilitating rapid agreement of pediatric investigation and study plans by both agencies, the proposal addresses the feasibility of developing multiple medicines for a rare disease in a reduced time frame and in a limited number of patients. There are monthly teleconferences of this cluster.

A third area is parallel EMA-FDA scientific advice, which can cover any area of product development. The procedure is voluntary at the request of the sponsor and looks to facilitate global development by allowing them to engage with both agencies at the same time. Following interactions between all three parties, and bilateral discussions between the two agencies, the EMA and FDA issue separate responses to the sponsor's questions in line with their usual procedures.

Cooperation during the evaluation of applications also operates through cluster groups, focused on specific therapeutic areas and drug types; for example, oncology, vaccines, blood products, and biosimilars. It allows for exchange of assessment reports and reviews and discussion of the benefit-risk of medicines prior to coming to a regulatory decision.

Once drugs are approved and used in patients, cooperation continues in a broader international context. The pharmacovigilance cluster, which comes together every month, brings together the EMA and the FDA with counterparts from Health Canada and Japan's PMDA/MHLW. In addition to the regular meetings, there are *ad hoc* exchanges of information on specific products and safety issues as they arise. The EMA operates an early notification system of upcoming safety issues that it also shares with Australia's TGA in addition to the cluster participants, which alerts regulatory partners to imminent regulatory action and public communication.

Cooperation in Inspections

With manufacturing and clinical trials increasingly taking place outside of the EU and US territory, cooperation on inspections has become a particularly important area for cooperation. In all instances, the objective is to better allocate inspection capacity, allowing more overseas facilities to be monitored and reduce unnecessary duplication.

Inspection programs exist today for "good manufacturing practice" (GMP) for active pharmaceutical ingredients and for finished products, and "good clinical practice" (GCP) inspections for both clinical trials and bioequivalence studies.

Given the international nature of inspections, the programs involve a number of regulators in addition to European regulators and the FDA, including those from Australia and the World Health Organization. Some of the programs focus on efficient distribution of inspection resources and work sharing; others on increased exchange of information on planned and completed inspection and increasing mutual reliance on each other's inspections.

In the area of GMP inspections, the EMA and FDA have joined forces, together with the EC, to create the EU/FDA Mutual Reliance Initiative in May 2014 aimed at reviewing each other's systems and moving forward in reliance on each other's inspection results; reducing duplication, making better use of resources, and increasing inspection coverage.

The FDA has attended joint audits of three EU member states as observers and is planning to observe another five audits in 2015. The EU will undertake audits of the FDA in September 2015. In addition, both sides have exchanged extensive information on each other's systems and compared legal frameworks. By the end of 2015, it is expected that the Mutual Reliance Initiative will have devised a process for sharing of unredacted inspection reports and for relying on each other's inspection results.

BENEFITS OF INTERNATIONAL COOPERATION

The past two decades have seen the pharmaceutical industry become truly global. Regulators have not been slow to recognize the need for cooperation with their international counterparts.

Globalized manufacturing of active ingredients and finished drugs, increasingly global clinical trials, interactions with international counterpart authorities, and the need for overseas inspections, all present a new set of challenges for regulators. The main current public health challenges are two-fold; how to ensure product quality and supply chain security, and how to ensure the integrity of data used to support clinical trials and manufacturing.

International cooperation seeks to focus on how to support a global approach to authorization and supervision of medicines in a risk-based approach, avoiding unnecessary duplication of efforts, creating synergies to ensure more effective use of global regulatory resources. Recent efforts such as the International Coalition of Medicines Regulatory Authorities (ICMRA) seek to go beyond this with a strategic and global approach to capacity building.

International cooperation provides for more efficient use of resources, in support of a risk-based approach, and contributes to increased consistency and predictability in regulatory approaches. It supports a global approach to development, authorization, and supervision of medicines and supports the availability of safe and effective new medicines, as early as possible, with data relevant to all regions.

The way in which international cooperation manifests itself has changed gradually. While there will always be a place for harmonization of certain rules and standards, there has been a shift to also include an emphasis on convergence, cooperation and work-sharing. It has also been characterized by more efforts in the area of capacity and competence building between authorities, including debate on regulatory science, and more recently work on developing comparability and assessment criteria to benchmark agencies and their work.

Many agencies, including the European Medicines Agency, have realized that it is no longer a question of whether to cooperate and collaborate with our partners: it is now about when and how best to do so. Only by taking this approach can we achieve our long-term goal, a truly global approach, to benefit patients both in Europe and throughout the world.

REFERENCES

[1] EMA website (about us), http://www.ema.europa.eu/ema/index.jsp?curl=pages/about_us/general/general_content_000235.jsp&mid=.

[2] Central authorization of medicines, http://www.ema.europa.eu/ema/index.jsp?curl=pages/about_us/general/general_content_000109.jsp&mid=WC0b01ac0580028a47.

[3] Human medicines: regulatory information, http://www.ema.europa.eu/ema/index.jsp?curl=pages/regulation/landing/human_medicines_regulatory.jsp&mid=.

[4] EMA guidance for companies requesting scientific advice or protocol assistance. Available online at: http://www.ema.europa.eu/ema/index.jsp?curl=pages/regulation/general/general_content_000049.jsp&mid=WC0b01ac05800229b9.

[5] CHMP Scientific Advice and Novel methods qualification (e.g., Biomarker), http://www.ema.europa.eu/ema/index.jsp?curl=pages/regulation/general/general_content_000049.jsp&mid=WC0b01ac05800229b9.

[6] EMA SME office, http://www.ema.europa.eu/ema/index.jsp?curl=pages/regulation/general/general_content_000059.jsp&mid=WC0b01ac05800240cc.

[7] SME 2014 annual report, http://www.ema.europa.eu/docs/en_GB/document_library/Report/2015/02/WC500181837.pdf.

[8] Patients involvement at EMA, http://www.ema.europa.eu/ema/index.jsp?curl=pages/partners_and_networks/general/general_content_000317.jsp&mid=WC0b01ac058003500c.

[9] Transparency, http://www.ema.europa.eu/ema/index.jsp?curl=pages/special_topics/general/general_content_000526.jsp&mid=WC0b01ac0580789730.

[10] EMA/ FDA cooperation, http://www.ema.europa.eu/ema/index.jsp?curl=pages/partners_and_networks/document_listing/document_listing_000228.jsp&mid=WC0b01ac058003176e.

Chapter 4

Human Clinical Trials and Drug Approvals: Transnational Issues

Ben Haas[*], Shira N. Epstien[*]
[*]*Latham & Watkins, LLP, Washington, District of Columbia, USA*

Chapter Outline

INTRODUCTION

To remain competitive, drug, device, and biotechnology companies must continually pursue innovative new products and product improvements. Clinical trials, which are essential to the development of new pharmaceutical, medical device, and biotechnology products, require the investment of tremendous time and resources—with no guarantee of success. In 2013, members of the Pharmaceutical Research and Manufacturers of America (PhRMA), the trade organization that represents the country's leading pharmaceutical research and biopharmaceutical companies, invested an

Food and Drug Regulation in an Era of Globalized Markets. http://dx.doi.org/10.1016/B978-0-12-802311-2.00004-7

estimated $51.1 billion in research and development (R&D),[1] with clinical trials typically consuming 40% or more of a typical firm's R&D budget.[2]

Clinical trials involve a wide range of participants, all of whom the sponsor is responsible for coordinating and overseeing. Major players include investigators, subjects, an institutional review board, and regulatory agencies, as well as other entities such as contract research organizations, drug manufacturers, insurance providers, researchers, scientists, and legal counsel. The overwhelming number of individuals involved in a clinical trial, as well as the increasing costs of clinical development and the length of the drug development process, which can take from 9 to 12 years from preclinical testing through Phase 3 clinical trials, have caused drug companies to explore opportunities to lower the costs and shorten the time lines of their clinical trial programs.

It therefore comes as no surprise that pharmaceutical companies in the United States have steadily increased the number of clinical trials they conduct overseas, with a focus on jurisdictions promising cost savings and recruitment efficiencies, such as those located in Central and Eastern Europe (CEE).[3] While U.S. companies can realize significant benefits by conducting a clinical trial in the CEE countries and other foreign jurisdictions, they must consider and account for the myriad legal, regulatory, ethical, and cultural questions they will invariably encounter in setting up and conducting a clinical trial in one or more foreign jurisdictions. In doing so, companies must retain a degree of flexibility, as both U.S. agencies and international bodies amend their clinical trial regulations from time to time. For example, in April 2008, FDA amended its foreign clinical trial regulations to ensure "the quality and integrity of foreign clinical data supporting FDA decision making on product applications and to help ensure the protection of human subjects participating in foreign clinical trials."[4] Additionally, in April 2014, the European Union (EU) adopted a new regulation that completely overhauls the clinical trial legal framework in the EU. Firms may have as little as 6 months to prepare for compliance with the new regulation before it goes into force.[5]

---

1. Pharmaceutical Research and Manufacturers of America, 2014. New Analysis Incorporates Real World Insights from Industry Leaders. Available at http://www.phrma.org/media-releases/economic-futures-report-release.

2. Kermani, F., 2006. CROs Stand to Benefit from Dramatic Rises in Pharma R&D Spending. Clinical Trials Today. Available at http://www.centerwatch.com/news-online/article/548#sthash.LlURoJ2b.dpbs.

3. The CEE includes, among others, Bulgaria, Croatia, Czech Republic, Hungary, Poland, Romania and Slovakia.

4. Getz, K., Zuckerman, R., 2008. Clinical Research Outsourcing: Moving from Transactional to Strategic Partnership-Based Outsourcing. Contract Pharma. Available at http://www.contractpharma.com/issues/2008-06/view_features/clinical-research-outsourcing/.

5. Wright, E., Farrell, C., 2014. New European Union Clinical Trials Regulation Adopted on 14 April 2014. Available at http://www.hlregulation.com/2014/04/15/new-european-union-clinical-trials-regulation-adopted-on-14-april-2014/.

TAKING CLINICAL TRIALS ABROAD

Enrolling patients in clinical trials in the United States has become increasingly slow, cumbersome, and frustrating to sponsors. Nearly 80% of clinical trials fail to meet their enrolment timelines and up to 50% of research sites enroll one or no patients. As of 2012, only 4% of Americans had ever been involved in a clinical trial.[6] Many potential volunteers are ineligible to participate in trials because they are often on multiple medications, making it difficult for sponsors to satisfy enrollment criteria and obtain usable data.[7] Faced with slow enrollment rates, increased competition for potential subjects, and rising patient and institutional costs, sponsors have set their sights abroad. From 2002 through 2009, the number of FDA-regulated investigators based outside the United States grew by 15% annually, whereas the number of US-based investigators declined by 5.5%. In addition, in 2007, approximately one-third of Phase 3 clinical trials for the 20 largest US-based pharmaceutical companies were conducted solely outside the United States.[8]

Over the last decade, the number of multicenter clinical trials conducted in CEE countries has grown at an average annual rate of 30% and this growth rate is expected to continue.[9] Sponsors are attracted to patient populations in CEE countries due to a number of factors: easier access to potential subjects; the ability to recruit patients more quickly from an expanded potential subject pool; lower overall trial costs; the availability of eager and qualified investigators; and the increased likelihood that enrolled subjects will complete a study. In addition, CEE countries offer sponsors access to subjects who are "clean" of confounding medications, and who do not otherwise have access to approved treatments for the disease being studied. The abundance of undertreated subjects in many of the CEE countries may facilitate rapid and steady enrollment, allowing sponsors to accelerate the recruitment process and obtain results more

---

6. Earls, E., 2012. Clinical Trial Delays: America's Patient Recruitment Dilemma. Drugdevelopment-technology.com. Available at http://www.drugdevelopment-technology.com/features/featureclinical-trial-patient-recruitment/.

7. Shah, S., 2002. Globalizing Clinical Research. The Nation. Available at http://www.thenation.com/article/globalizing-clinical-research.

8. Glickman, S. W., McHutchison, J. G., Peterson, E. D., Cairns, C. B., Harrington, R. A., Califf, R. M., Schulman, K. A., 2009. Ethical and Scientific Implications of the Globalization of Clinical Research. The New England Journal of Medicine 360, 816-823.

9. Barnes, K., 2006. CROs March Towards Eastern Europe. OutSourcingPharma. Available at http://www.outsourcing-pharma.com/Clinical-Development/CROs-march-towards-Eastern-Europe (At 2011 retail prices, the pharmaceutical market in CEE countries was estimated to be $63.6 billion. It has grown rapidly over the past few years and is expected to reach around $104.2 billion by 2016). Bombourg, N., 2012. The Outlook for Pharmaceuticals in Central & Eastern Europe. PR Newswire. Available at http://www.prnewswire.com/news-releases/the-outlook-for-pharmaceuticals-in-central--eastern-europe-183334411.html. *See also* Novack, T., Belotserkovsky, M., Fekete, V., Moscicka, K., 2013. Clinical Trial Landscape in Central and Eastern Europe. Clinical Informatics News. Available at http://www.clinicalinformaticsnews.com/cln/2013/9/12/clinical-trial-landscape-central-eastern-europe.html.

quickly. Further, the CEE is becoming increasingly attractive as the FDA and other regulatory authorities worldwide participate in efforts to harmonize international GCPs.

However, U.S. companies that are considering sponsoring clinical trials at foreign sites need to be aware that not all CEE countries have fully implemented EU rules and regulations governing clinical trials and data privacy.[10] This has resulted in clinical trial regulations that differ from country to country, often leading to confusion for sponsors who are conducting multinational trials. While the EU's recently adopted Clinical Trials Regulation represents the European Commission's attempt to harmonize some aspects of the legal framework surrounding clinical trials, the new regulation leaves several aspects of clinical trials governance to individual member nations.

Not only do sponsors face variations in the legal and regulatory requirements from country to country, they also face varying and perplexing local regulations, cultures, and traditions, as well as the obvious barrier of communicating in different languages. In addition, many developing countries lack the governmental oversight, resources, and regulatory structure to ensure that sites and investigators treat subjects according to recognized ethical standards, leaving sponsors with the responsibility of overseeing the details of patient enrollment, disclosures, and treatment during the foreign trials to ensure appropriate and ethical conduct.[11] These variances among the requirements and expectations in each country in which a study may be conducted can cause delays in the start of trials and may impede a sponsor's ability to meet time lines for completion. They may also increase costs as sponsors struggle to become familiar with local regulations, the requirements and expectations of local institutions and investigators, and the need to address local issues in a way that protects the sponsors' interests in confidentiality and intellectual property and ensures that the data produced by the sites will satisfy the requirements and expectations of the FDA and other nations' regulatory agencies.

OVERLAPPING LEGAL AND REGULATORY REQUIREMENTS

Legal Landscape in the United States: FDA Oversight of Clinical Trials

The FDA extensively regulates the use, marketing, and testing of new drugs and biologics in the United States. To do so, the FDA promulgates its own regulations under the Federal Food, Drug, and Cosmetic Act (FDCA). These rules

10. Ward, P., 2006. European Clinical Trial Regulations Undergo a Slow Revolution. Pharmaceutical Regulatory Guidance Book. Available at http://www.pharmtech.com/pharmtech/data/articlestandard/lcgc/142007/416308/article.pdf.

11. Concar, D., 2003. A Bitter Pill: Exporting Drug Trials to the Developing World. Amnesty International Magazine. Available at www.amnestyusa.org/page.do?id=1105438.

apply to clinical trials conducted to support U.S. approval, regardless of where the study is conducted. In order to market a new drug in the United States, a company must obtain FDA approval of a New Drug Application (NDA). To obtain NDA approval, the sponsor must demonstrate, to the FDA's satisfaction, that the drug is safe and effective for its intended use(s) through the generation and submission of adequate and well-controlled clinical data. Thus, a sponsor must conduct a series of clinical trials on human subjects, generally at multiple sites, either in the United States or abroad. Typically, clinical trials are conducted in three sequential phases, which may overlap.[12] Phase 1 trials are conducted on a limited number of (usually healthy) subjects to test the absorption, dose tolerance, safety, metabolism, distribution, and excretion of the drug candidate; Phase 2 trials are conducted on larger subject groups to identify possible adverse effects and safety risks and to determine dose tolerance, optimal dosage, and the efficacy of the drug candidate for specific targeted indications; and Phase 3 studies are conducted on large and diverse subject populations at multiple, geographically dispersed clinical trial sites, and are undertaken to further evaluate dosage, assess clinical efficacy, and evaluate safety.[13] Phase 3 trials are commonly referred to as pivotal trials, and the data obtained from these studies are used by the sponsor to support its NDA application to the FDA for approval of the drug candidate.

To distribute an unapproved new drug in interstate commerce for clinical testing in the United States, the sponsor must submit an investigational new drug (IND) application to FDA.[14] The IND application must contain animal data, relevant literature, and available human data demonstrating that the drug candidate is safe for human testing. The sponsor must also submit the clinical trial protocol(s) for the planned study or studies and information related to the chemistry and manufacture of the drug candidate.[15] The IND goes into effect 30 days after receipt by the FDA, unless the FDA places the trial on clinical hold.[16] Once the IND becomes effective, the sponsor may commence its clinical trial consistent with the FDA's good clinical practice (GCP) regulations, which require, among other things, patient informed consent, oversight by an institutional review board (IRB), investigator statements, and adverse event and periodic reporting to the FDA.[17]

All clinical research conducted in the United States in support of an NDA must comply with the FDA's IND regulations. Trials that are conducted outside of the United States to support an NDA may, but need not, comply with these requirements. In light of the significant increase in clinical trials conducted

12. 21 Code of Federal Regulations § 312.21.
13. 21 Code of Federal Regulations § 312.21(c).
14. 21 Code of Federal Regulations § 312.
15. 21 Code of Federal Regulations § 312.23.
16. 21 Code of Federal Regulations § 312.40.
17. 21 Code of Federal Regulations § 312(D).

overseas, the FDA has made several changes to its regulations to improve its ability to verify the quality of foreign data and ensure that subjects enrolled in foreign clinical trials are adequately protected. For example, the Food and Drug Administration Amendments Act of 2007 introduced a requirement for the registration of almost all clinical trials (except Phase 1 trials) and increased the FDA's ability to require postmarket studies.[18] Additionally, in 2008, the FDA amended the regulations governing the use of foreign trials that are not conducted under an IND.[19] In amending the regulations, the FDA modified the standards under which the agency will accept data from foreign clinical trials in support of domestic applications and submissions to require, among other things, that the trials be conducted in compliance with the GCP requirements.[20] The FDA believes that requiring compliance with the GCP standards will ensure adequate protection of human subjects while providing the flexibility necessary to accommodate differences in how countries regulate clinical research and obtain informed consent.[21]

Before the FDA's amendment to the foreign clinical trial regulations, data from a non-IND foreign clinical study would be accepted and reviewed by the FDA in connection with an NDA submission if the study was conducted in compliance with the ethical standards of the Declaration of Helsinki (an international guideline for conducting human subject research) and/or the laws and regulations of the country or countries in which the study was performed, whichever offers the greater protection to human subjects.[22] Under the amended foreign clinical trial regulations, in addition to complying with the national and local regulations imposed by the host country (in which the trial is being conducted), sponsors must also demonstrate that the studies are conducted in accordance with GCP, and must permit the FDA to validate the data through onsite inspection.[23] The GCP regulations also require that the study be conducted under the oversight of an Independent Ethics Committee (IEC)—"a review panel that is responsible for ensuring the protection of the rights, safety, and well-being of human subjects involved in a clinical investigation and is adequately constituted to provide assurance of that protection."[24] The amended regulations apply to all foreign clinical studies regardless of the status of subject enrollment, whether ongoing, completed, or not yet initiated.

Clinical trials conducted in the United States are also subject to the Health Insurance Portability and Accountability Act (HIPAA), which, among other things, provides important privacy and security safeguards to protect the confidentiality of patients' personal health information. The HIPAA privacy and

18. H. R. 3580, 110th Congress, 2007. Food and Drug Administration Amendments Act of 2007.
19. 73 Federal Register § 22, 800.
20. 21 Code of Federal Regulations § 312.120.
21. 73 Federal Register § 22, 800.
22. 21 Code of Federal Regulations § 312.120.
23. 21 Code of Federal Regulations § 312.120(a).
24. 21 Code of Federal Regulations § 312.3.

security regulations establish complex and comprehensive federal standards relating to the use and disclosure of protected health information and set minimum standards to protect the confidentiality and availability of patients' health information.[25] In addition, a number of states have enacted laws relating to the confidentiality of patient's medical information.[26] The HIPAA regulations do not supersede more stringent state laws; therefore, sponsors are required to comply with both federal privacy regulations and varying state privacy laws.

Legal Landscape in the EU: EU Clinical Trial Directives and Regulation

In May 2004, the EU required member states to implement the EU Clinical Trials Directive 2001/20/EC (CTD), which was intended to simplify and harmonize the requirements for conducting clinical trials within the EU. In general, the CTD established minimum standards for conducting Phase 1 through 4 clinical trials, and was intended to ensure that human studies are conducted ethically, safely, credibly, and efficiently. The CTD provided the statutory basis for GCP standards for clinical trials and good manufacturing practice (GMP) standards for the drug used in those trials. The CTD also established standards relating to the organization and oversight authority of ethics committees, protection of clinical subjects (including informed consent, confidentiality of patient personal information, and risk/benefit analysis), labeling, suspected unexpected serious adverse reactions (SUSAR) reporting and monitoring, and harmonization of the application procedure. A key result of implementation of the CTD was the development of the EudraCT database, which allows EU regulators to share information on clinical trial submissions, amendments, ethical approvals, favorable ethics committees opinions, suspected SUSARS, and GCP inspections.[27]

In April 2005, the EU implemented Directive 2005/28/EC, which establishes additional GCP requirements, including standards for the design, conduct, recording, and reporting of clinical trials (including the requirement that clinical trials be conducted in accordance with the Declaration of Helsinki), standards and requirements relating to the authorization of the manufacture and importation of investigational drug products, and guidelines on the documentation relating to clinical trials, the qualifications of inspectors, and inspection procedures and archiving.[28] Directive 2003/94/EC on GMP establishes standards for the

25. 45 Code of Federal Regulations §§ 160, 164.
26. California Civil Code § 56.10.
27. Official Journal of the EU, 2003. Commission Directive 2003/94/EC of 8 October 2003. Available at http://ec.europa.eu/enterprise/pharmaceuticals/eudralex/vol-1/dir_2003_94/dir_2003_94_en.pdf; European Medicines Agency, Inspections—Good Manufacturing Practice. Available at http://www.emea.europa.eu/Inspections/GMPhome.html.
28. Official Journal of the EU, 2005. Commission Directive 2005/28/EC of 8 April 2005. Available at http://europa.eu.int/eurlex/lex/LexUriServ/site/en/oj/2005/l_091/l_09120050409en00130019.pdf.

manufacture of the investigational drug, inspections, quality assurance systems, documentation requirements, quality control systems, complaints, recalls, and product labeling.[29]

In April 2014, the EU adopted a new Regulation of the European Parliament and of the Council on Clinical Trials on Medicinal Products for Human Use, and repealing Directive 2001/20/EC Clinical Trials Regulation.[30] The new Clinical Trials Regulation (CTR) will replace the existing Clinical Trials Directive (CTD) and will be directly binding in all EU member states without the need for national implementing legislation. The CTR completely overhauls the existing legal framework surrounding clinical trials in Europe, a welcome development in light of the long-standing criticism of the Clinical Trials Directive as unwieldy and inefficient. According to a study ordered by the European Commission (EC), the number of clinical trial applications in the EU has fallen drastically since the CTD came into force (between 2003 and 2007, the number fell by 25%), the administrative cost of conducting a clinical trial in the EU has risen dramatically, and the staffing required by commercial sponsors has more than doubled.[31]

The Clinical Trials Regulation implements several changes to address concerns associated with the CTD. The CTR reduces the number of categories of clinical trial postauthorization modifications requiring approval; streamlines the clinical trial authorization process by providing for a harmonized application dossier and a clear-cut two-part application assessment; creates a less burdensome pathway for low-intervention clinical trials; and introduces cosponsorship of trials to reflect the current practice of various pharmaceutical companies and/or research institutions joining forces to conduct clinical trials.[32] These and the other changes in the Clinical Trials Regulation constitute a renewed attempt by EU institutions to harmonize and simplify the clinical trial legal framework after the CTD's failings. However, the CTR maintains a significant degree of member state autonomy with respect to some important aspects, such as investigator and subject compensation and recruitment, and the assessment of the suitability of investigators and sites—areas where differences in standards and practices between member states are perhaps more acute.

---

29. Official Journal of the EU, 2003. Commission Directive 2003/94/EC of 8 October 2003. Available at http://ec.europa.eu/enterprise/pharmaceuticals/eudralex/vol-1/dir_2003_94/dir_2003_94_en.pdf; European Medicines Agency, Inspections—Good Manufacturing Practice. Available at http://www.emea.europa.eu/Inspections/GMPhome.html.

30. Regulation (EU) No 536/2014 of the European Parliament and of the Council of 16 April 2014 on clinical trials on medicinal products for human use, and repealing Directive 2001/20/EC, 2014 O.J. (L158).

31. Impact on Clinical Research of European Legislation (ICREL), Project Final Report. Available at http://www.efgcp.be/downloads/icrel_docs/Final_report_ICREL.pdf.

32. Regulation (EU) No 536/2014 of the European Parliament and of the Council of 16 April 2014 on clinical trials on medicinal products for human use, and repealing Directive 2001/20/EC, 2014 O.J. (L158).

International Legal Landscape: International Clinical Practice Standards

The Declaration of Helsinki was developed in 1964 by the World Medical Association and has undergone several revisions (1975, 1983, 1989, 1996, 2000, 2008, and in 2013). It provides a set of ethical guidelines for physicians and other participants in medical research. These guidelines require, among other things, that clinical investigators be aware of the ethical, legal, and regulatory requirements for research on human subjects both in their own countries and internationally, and that no national, ethical, legal, or regulatory requirement may reduce or eliminate any of the protections set forth in the Declaration of Helsinki. The Declaration of Helsinki includes principles on minimizing risk, confidentiality of patient information, informed consent, conflicts of interest, research protocol, and safeguarding participants in research.[33]

The International Conference on Harmonization Good Clinical Practice Guidelines (ICH GCP) sets forth international standards for the quality, safety, and efficacy of developmental-stage pharmaceutical products. The ICH GCP were made binding by the implementation of the EU Clinical Trials Directive in 2004 and remain so with the new Clinical Trials Regulation. They require the sponsor to verify the qualifications of the investigator(s), obtain written informed consent before each subject's participation in the trial, ensure the trials are adequately monitored, and that the appropriate institutional review boards adequately review and approve the protocol as well as oversee the trial. The ICH GCP guidelines specify the responsibilities of the IRB, including review of recruitment advertisements, the informed consent document, and subject compensation.

National and Local Legal Standards

Finally, a sponsor undertaking a clinical trial abroad must adhere to national or local requirements of the country or countries in which the study is conducted. Depending on the particular countries at issue, these may vary in terms of submission contents, time lines, ongoing oversight, and reporting obligations. Additionally, some countries that are not part of the EU may have wholly independent clinical trial regulations and processes.

PRACTICAL CONSIDERATIONS IN CONDUCTING GLOBAL TRIALS

In order to obtain marketing approval in the United States, a sponsor must obtain data from clinical trials that demonstrate that the sponsor's drug is safe and effective for the drug's proposed use(s) and sufficient to support marketing approval. As part of doing so, the sponsor should obtain early regulatory feedback

33. World Medical Association, 2004. Declaration of Helsinki. Available at http://www.wma.net/en/30publications/10policies/b3/.

on clinical trial design and analysis, and must comply with GCP regulations. The clinical development process is costly and burdensome, and a sponsor must carefully consider whether conducting a clinical trial in whole or in part overseas will provide a cost-effective and efficient pathway to generating clinical data acceptable to FDA. Additionally, sponsors must meet various regulatory objectives including protecting research subjects, ensuring research is scientifically sound, avoiding fraud and bias, and ensuring the quality and purity of the investigational product. After expending such significant effort to develop the clinical trial, the sponsor will also want to protect his or her investment with confidentiality safeguards, intellectual property rights, and consolidated publication from all clinical trial sites.

The following are some key considerations and issues that should be addressed by a sponsor before choosing to conduct a foreign clinical study.

Deciding Whether to Conduct the Foreign Clinical Trial Under an IND

All clinical research conducted in the United States in support of an NDA or Biologics License Application (BLA), used for seeking approval for biological products, must comply with the FDA's IND regulations. A clinical trial that includes both United States and foreign trial sites will require an effective IND. However, sponsors who choose to initiate a trial in which all the clinical sites will be located abroad have the option of submitting an IND to the FDA, but need not do so. Submission of an IND for the study will result in FDA review of the protocol and the assurance that the FDA will remain in communication with the sponsor during the development process. However, obtaining an effective IND for a foreign clinical study will also result in an additional layer of regulation, which can be confusing and burdensome—particularly to foreign sites and investigators who lack familiarity or experience with the U.S. regulations. The FDA's amended regulations applicable to non-IND foreign trials now require that such trials be conducted pursuant to GCP criteria, under IEC oversight and subject to FDA inspection, thus eliminating some of the flexibility previously associated with non-IND foreign trials.

Identifying Experienced Contractors and Contract Research Organizations

US-based sponsors of overseas trials are spending an increasing amount of their developmental budget on Contract Research Organizations (CROs), to help aid in the management and oversight of the study.[34] A sponsor may transfer some,

34. Getz, K., Zuckerman, R., 2008. Clinical Research Outsourcing: Moving from Transactional to Strategic Partnership-Based Outsourcing. Contract Pharma. Available at http://www.contractpharma.com/issues/2008-06/view_features/clinical-research-outsourcing/ (Sponsors spent 17 percent of their total development budget on CRO services in 2007, up from 14 percent in 2001).

or all, of its responsibilities in conducting clinical trials under the FDA's regulations to a CRO.[35] CROs that have experience in the local countries and speak the local languages can have a significant impact on the timely and efficient performance of the study. As a result, sponsors often pay millions of dollars to CROs in connection with large Phase 3 studies. However, ceding responsibility for the timely, cost-efficient, and lawful performance of a foreign clinical study entirely to a third party entails significant risks. Most importantly, a CRO is a service provider with multiple clients, and thus may not share the sponsor's commitment to the study. Entering into an arrangement with a foreign CRO raises many of the same issues faced by sponsors when considering foreign sites and investigators, including language barriers, familiarity with local requirements and resources, problem solving skills, and knowledge of U.S. regulatory requirements.

As a result, when considering CROs, a sponsor should not necessarily look for the "best offer" financially, but instead should focus on the CRO that offers significant experience, will devote adequate resources, and is highly knowledgeable about local laws, sites, and investigators. Further, in contracting with a CRO, a sponsor should generally maintain control over the project team to avoid substitution with less-qualified personnel once the contract is signed and the study has commenced, leaving the sponsor with few alternatives. Sponsors should also commit the time and effort to ensure that the contract specifically describes the services to be performed by the CRO and the associated budget. This will minimize unexpected fees and budget increases after-the-fact for particular services or requests that the CRO deems to be "out-of-scope."

Managing Idiosyncratic Foreign Sites and Investigators

Sponsors are often frustrated by the time and effort it takes to retain foreign sites and investigators on terms acceptable to the parties. Sites and investigators, particularly those located in the CEE, often raise esoteric and idiosyncratic requests and issues when negotiating and executing clinical trial agreements and in addressing other aspects of the clinical trial. For instance, the site may request that the sponsor enter into a separate agreement with each entity involved in the study at the site, including investigators, subinvestigators, pharmacies, and study nurses, in order to allow for individualized budgets and payments. Similarly, a foreign investigator may request a separate agreement with the sponsor, so that the investigator may receive payment directly from the sponsor without going through the site. Such requests raise important legal and practical issues that must be addressed by the sponsor, and may delay contract negotiation and execution.

Additionally, sponsors' and investigators' contract goals may be at odds with each other. Sponsors prioritize quality control, protocol, and regulatory compliance, confidentiality, protection of intellectual property, ownership of

35. 21 Code of Federal Regulations § 312.52.

data and results, budget control, indemnification, and IND compliance, where applicable. Investigators and their corresponding clinical trial sites are more concerned about getting paid, publication rights, and local requirements governing clinical trials. Drafting an agreement that meets the needs of both sponsors and investigators requires significant time and effort, and both parties should strive for consistent key contract terms across sites.

Other common issues that arise in negotiations with foreign sites and investigators include: (1) the laws and language that will govern the agreement; (2) ensuring proper translations of the agreement; (3) the availability, scope, and coverage of indemnification for the sites' and investigators' negligence or misconduct; (4) the adequacy of the sites' insurance coverage and compliance with local insurance requirements; (5) protection of intellectual property and confidentiality; (6) ethical considerations relating to the human subjects; and (7) data publication rights. Careful and early consideration of these issues, and ensuring that the clinical trial agreement accurately reflects the parties' understanding of how the issues will be handled, will help mitigate the risk of disputes or miscommunications after the trial has commenced.

CONCLUSION

Clinical trials are increasingly global in scope. Global trials offer a larger patient pool, which facilitates more rapid enrollment, and offer lower fees and administrative costs. Additionally, global trials may offer more eager and qualified investigators and a greater likelihood of completing a study within established time lines. Yet, while conducting a clinical trial overseas can benefit sponsors, it is also fraught with potential risks and pitfalls. A sponsor may face increased competition, resulting in fewer available clinical trial sites, as well as increased costs, travel, contracting requirements, and delays. A sponsor can mitigate these risks by considering the following suggestions:

A. *Start Early*

A sponsor should anticipate "bumps in the road" when conducting trials in foreign jurisdictions. Compared to conducting studies in the United States, setting up clinical trials overseas will almost assuredly require the investment of more time, effort, and resources. However, if done correctly, the upfront investment may be offset by the rapid and steady enrollment of subjects and the collection of data once the study is commenced.

B. *Ensure Regulatory Compliance*

With extensive and potentially overlapping regulatory requirements that may be imposed on a multinational trial, sponsors should take steps, early in the process, to assess their strategic goals and options and to identify and develop compliance plans that address all applicable FDA, EU, international, national, and local laws.

C. *Focus on Clinical Trial Agreements*

Clinical trial time lines may be unexpectedly delayed due to a failure to execute the agreements necessary to allow sites and investigators to participate in the trial. Sponsors should be alert to potential areas of conflict and pay close attention to the drafting and negotiation of clinical trial agreements, even if a CRO for the study has primary responsibility for the contracts.

D. *Steer Clear of Inexperienced CROs*

Sponsors should seek CROs that have experience in the local jurisdictions and have successfully identified and managed sites and investigators in the relevant countries. The premium of such experience often outweighs the lower cost that may be offered by a less-experienced and less-qualified CRO.

E. *Maintain Control*

Sponsors should not cede the right to manage and control the clinical study to a third-party vendor. In particular, a sponsor should consider maintaining control over the study through regular updates and meetings with CROs, and retaining contractual authority over the study team. Sponsors should retain the right to review and approve changes to the sites, investigators, and study agreements, to prevent problems or changes from occurring without the sponsor's knowledge.

FURTHER READING

European Commission, 2004. Detailed Guidance on the European Clinical Trials Database (EUDRACT Database). Available at http://ec.europa.eu/enterprise/pharmaceuticals/eudralex/vol10/13_cp_and_guidance:eudract_april_04.pdf.

European Commission Enterprise Directorate-General, 2004. Amendment Describing Deployment of EudraCT- Lot 1 for 1 May 2004. Available at http://www.meduni-graz.at/ethikkommission/Forum/Download/Files/CT_5_1.pdf.

73 Federal Register 22, 800, 22, 813. Human Subject Protection: Foreign Clinical Studies Not Conducted Under an Investigational New Drug Application.

International Conference on Harmonisation of Technical Requirements for Registration of Pharmaceuticals for Human Use, 1996. ICH Harmonised Tripartite Guideline—Guideline for Good Clinical Practice E6(R1). Available at http://www.ich.org/fileadmin/Public_Web_Site/ICH_Products/Guidelines/Efficacy/E6/E6_R1_Guideline.pdf.

Chapter 5

Falsified and Substandard Medicines in Globalized Pharmaceutical Supply Chains: Toward Actionable Solutions

Sam F. Halabi*,[†], Lawrence O. Gostin[†]
*The University of Tulsa College of Law and Scholar, Tulsa, Oklahoma, USA, [†]O'Neill Institute for National and Global Health Law, Washington, District of Columbia, USA

Chapter Outline

Globalization—of product supply chains, labor, finance, crime, and regulation—generates both opportunities to raise the global standard of human quality of life as well as deep, vexing challenges in securing what gains have been achieved from criminal and illicit networks that now operate more widely than ever before.[1] The human right to health, particularly that aspect of the right that involves access to essential medicines, exemplifies these countercurrents. On the one hand, globalization has facilitated access to medicines for more of the world's population than ever before. On the other hand, the worldwide reach of pharmaceutical supply chains has opened

1. Wolf, M., 2005. Why Globalization Works, second ed. Yale University Press; Stiglitz, J., 2003. Globalization and Its Discontents. W. W. Norton & Company, London.

Food and Drug Regulation in an Era of Globalized Markets. http://dx.doi.org/10.1016/B978-0-12-802311-2.00005-9

up numerous points of entry for the introduction of counterfeit, adulterated, falsified, or substandard medicines. These phenomena are not equal and opposing. Delivering diluted or substandard antibiotics, for example, causes mass selection of antibiotic-resistant microbes, resulting in harm even to those populations who receive adequate and effective antibiotic courses. Moreover, the problem of falsified and substandard medicines disproportionately impacts the poor in South and Southeast Asia and sub-Saharan Africa. In this chapter, we analyze the problem of falsified and substandard medicines as a globalization problem that requires a globalized solution. National regulatory authorities must forge common terms, understandings, and technological approaches including as critical matters areas in which investments must be made in low- and middle-income regulatory capacity and whether counterfeit medicines are approached principally as an intellectual property or a public health problem.[2]

GLOBALIZATION'S INFLUENCE ON THE PHARMACEUTICAL SUPPLY CHAIN

Every aspect of the pharmaceutical supply chain has increasingly migrated to low- and middle-income countries or expanded where it had already existed. The result is increased complexity in the global pharmaceutical supply chain and an expansion of entry points for adulterated, negligently manufactured, and fraudulently labeled medicines. For example, a US-finished drug may be produced using an active pharmaceutical ingredient (API) made in China and ingredients made in Europe, Japan, or the United States. These components may be shipped to India where the finished drug is manufactured and then imported into the United States for distribution. Packaging and repackaging may take place in other countries. The further away from developed economies the supply chain moves, the more fragmented and difficult-to-regulate it becomes. Regulatory agencies—which must register medicines, issue licenses and marketing authorizations, monitor their populations for adverse events after a medicine has been approved, oversee clinical trials, and inspect facilities—are *always* tasked with more than their resources allow. In low- and middle-income countries these constraints are even tighter because of staffing shortages and low pay. A few national firms command most of the primary wholesale market in wealthy countries but in low- and middle-income countries hundreds or even thousands of firms participate. In some major developing economies like China and India, similar fragmentation obtains for manufacturing as well. Governmental and nongovernmental estimates of licensed pharmaceutical manufacturers in India

2. Clift, C., 2010. Combating counterfeit falsified, and substandard medicines: Defining the way forward? Chatham House Briefing Paper.

range from 5,877 to over 20,000, some of which are quite small.[3] For any given country, managing this sprawling supply chain is a "regulatory nightmare."[4]

From the beginning point in the pharmaceutical production process—APIs and the excipients in which they are suspended—to the point of sale or receipt by consumers and patients, globalization has greatly increased price volatility and opportunity for arbitrageurs, both licit and illicit, to exploit that volatility for profit. APIs may cost thousands of dollars per kilogram. The price increases, and may even double, as greater assurances of quality, like WHO prequalification or approval by a stringent regulatory authority, are obtained.[5] The market for active ingredients has become particularly volatile in recent years because of increasing costs of raw materials and expanding regulation in India and China.[6] China supplies about 43% of the world's APIs for anti-infective medicines and exports 77% of the active ingredient made in the country, a $4.4 billion industry.[7] India exports 75% of the $2 billion worth of the APIs it produces. APIs made in facilities lacking adequate air and water quality control are less expensive as, of course, are fake APIs.

Formulation companies combine APIs and excipient products to produce final medications that are in turn sold to a diverse set of wholesale purchasers. At the formulation stage, medicines may be diluted, negligently labeled, or fraudulently labeled. Warren Kaplan and Richard Laing found one-tenth of Indian registered pharmacies report substandard medicines, with a disproportionate number originating with small- and medium-sized producers.[8] Even if quality controlled, safe medicines are produced at the formulation stage, drug wholesalers persistently buy and sell medicines to meet demand. When they observe a drug scarcity in one region, they purchase that medicine from other wholesalers who have too much. The product flows constantly fluctuate and medicines change hands many times.[9] Wholesalers frequently repackage products with

3. KPMG International, 2006. The Indian pharmaceutical industry: Collaboration for Growth. Available at https://www.in.kpmg.com/pdf/Indian%20Pharma%20Outlook.pdf.

4. 2012. Bad medicine. The Economist. Available at http://www.economist.com/node/21564546.

5. Bumpas, J., Betsch, E., 2009. Exploratory study on active pharmaceutical ingredient manufacturing for essential medicines. Washington, DC: The International Bank for Reconstruction and Development/The World Bank.

6. Bumpas, J., Betsch, E., 2009. Exploratory study on active pharmaceutical ingredient manufacturing for essential medicines. Washington, DC: The International Bank for Reconstruction and Development/The World Bank.

7. Bumpas, J., Betsch, E., 2009. Exploratory study on active pharmaceutical ingredient manufacturing for essential medicines. Washington, DC: The International Bank for Reconstruction and Development/The World Bank.

8. Kaplan, W., Laing, R., 2005. Local production of pharmaceuticals: Industrial policy and access to medicines—an overview of key concepts, issues and opportunities for future research. Washington, DC: The International Bank for Reconstruction and Development/The World Bank.

9. Buckley G., Gostin, L.O., eds., Committee on Understanding the Global Public Health Implications of Substandard, Falsified, and Counterfeit Medical Products; Board on Global Health; Institute of Medicine. 2013. Countering the Problem of Falsified and Substandard Drugs. 197. National Academies Press, Washington, D.C. (hereinafter IOM Countering Falsified and Substandard Drugs).

each sale, or repackage containers for final sale.[10] "Through a process called salting, legitimate and fake drugs are mixed at wholesale, and in the wholesale repackaging, the fake products gain authentic labels."[11]

The economics of drug manufacturing increase the number of arm's-length and sometimes blind links in the supply chain.[12] Cost calculations drive both innovator and generic pharmaceutical companies to contract with manufacturers abroad. For example, establishing a pharmaceutical factory in India costs only 40% of establishing a similar facility in North America or Europe.[13] Firms supply contract manufacturers with the materials, including packaging, and accept material losses and other deviations as the cost of doing business. Dilip Shah, Secretary General of the Indian Pharmaceutical Alliance, explained:

> *Very few companies, foreign or domestic, monitor the [contract manufacturer's] process loss of raw materials, active ingredients, and packaging materials. I have known of cases of 15 to 20 percent packaging material losses and companies are not bothered.*[14]

Because contract manufacturers have established distribution channels, they are well positioned to introduce fake or substandard drugs into the distribution system. Because the contract manufacturers have the processes and materials needed to produce real medicines, they will sometimes sell drugs outside of the legal distribution system, using authentic packaging to launder their products.[15]

THE EFFECTS OF THE GLOBAL MARKET FOR FALSIFIED AND SUBSTANDARD MEDICINES

Unfortunately, there are currently no accurate estimates of the global burden of falsified and substandard drugs precisely because activity occurs in global black and gray markets. Evidence suggests, however, that the problem is most severe in low- and middle-income countries.[16] Individual and organized criminals exploit the relatively weak drug regulatory systems there, knowing that the

---

10. Catizone, C. A., 2006. Counterfeit drugs and states' efforts to combat the problem. Journal of Pharmacy Practice 19, 165-170.
11. IOM Countering Falsified and Substandard Drugs 204 (citing Donaldson, A., 2010. Current development: "And the ones that mother gives you don't do anything at all" combating counterfeit pharmaceuticals: The American and British perspectives. New England Journal of International and Comparative Law 16, 145-168; Liang, B. A., 2006. Fade to black: Importation and counterfeit drugs. American Journal of Law and Medicine 32, 279-323).
12. Price Waterhouse Cooper, 2010. Global pharma looks to India: Prospects for growth. Retrieved from http://www.pwc.com/gx/en/pharma-life-sciences/publications/india-growth.jhtml.
13. Price Waterhouse Cooper, 2010. Global pharma looks to India: Prospects for growth. Retrieved from http://www.pwc.com/gx/en/pharma-life-sciences/publications/india-growth.jhtml.
14. IOM Countering Falsified and Substandard Drugs, 146.
15. IOM Countering Falsified and Substandard Drugs, 120.
16. South African Development Community, 2007. Guidelines for submitting application for registration of medicine. Available at http://medicines.sadc.int/?page_id=119.

regulators are poorly equipped to handle the surveillance and enforcement measures that would be necessary to thwart illicit manufacturing and trade in falsified drugs.[17] The simplest operations use a manual press to load any powder into empty capsules bought on the open market. The legal and illegal supply chains mix in poorly regulated markets.[18] For example, the World Health Organization (WHO) found about 1000 unregistered drugs on the Cambodian market, which is historically a country where legal and illegal supply chains mix for distribution of falsified and nonfalsified antibiotics and antimalarial medications.[19] There is a second consequence for low- and middle-income countries afflicted by falsified drugs: the cost of medicine accounts for 20%-60% of health spending there, and 90% of the population pays for medicine out-of-pocket.[20]

Sometimes falsified and substandard medicines effectively poison consumers and patients. In 2008, overly sulfated chondroitin in the anticoagulant heparin likely killed 81 people and sickened hundreds more. Between 2008 and 2009, 84 Nigerian children died from kidney failure caused by the industrial solvent diethylene glycol in teething syrup.[21] Inspectors traced the problem to fraud committed by a chemical dealer in Lagos.[22] A similar tragedy occurred in Panama when a Chinese chemical manufacturer sold diethylene glycol, the active ingredient in antifreeze, as pharmaceutical-grade glycerin to a European firm.[23] The contamination killed 219 people.[24] In 2012, the Chinese State Food and Drug Administration (SFDA) found 13% of capsule manufacturers produced drugs containing unsafe levels of chromium.[25] The SFDA identified 254 separate companies as sources of the chromium-tainted medicines.[26]

Though poisonous drugs are a particularly frightening aspect of the global problem, the broader and deeper effect is medicine that simply does not work, especially for diseases that disproportionately affect low- and middle-income countries. In a convenience sample of pharmacies in Lagos, Nigeria, researchers found that four of eight popular brands of metformin tablets (a common

17. IOM Countering Falsified and Substandard Drugs, 154.
18. Clark, E., 2008. Counterfeit medicines: The pills that kill. The Telegraph.
19. Lepakhin, A., 2003. WHO launches campaign against counterfeit medicines. Bulletin of the World Health Organization 81, 921-922.
20. Cameron, A., Ewen, M., Ross-Degnan, D., Ball, D., Laing, R., 2008. Medicines prices, availability, and affordability in 36 developing and middle-income countries: A secondary analysis. The Lancet, 373, 240-249.
21. Polgreen, L., 2009. 84 children are killed by medicine in Nigeria. The New York Times.
22. Polgreen, L., 2009. 84 children are killed by medicine in Nigeria. The New York Times.
23. Bogdanich, W., Hooker, J., 2007. From China to Panama, a trail of poisoned medicine. The New York Times.
24. Núñez, E., 2011. Institute of medicine advances in legal review of pending cases: 219 are killed by syrup. Prensa.
25. Rickman, J., 2012. China shutters shoddy drugmakes in wake of chromium capsule scare. International Pharmaceutical Regulatory Monitor 40(6).
26. Rickman, J., 2012. China shutters shoddy drugmakes in wake of chromium capsule scare. International Pharmaceutical Regulatory Monitor 40(6).

diabetes treatment) failed one or more pharmacopeial tests of bioequivalence.[27] A Rwandan study found that 20% of medicines in a sample of pharmacies were substandard at the time of purchase.[28] In the 1990s, antibiotics accounted for over 45% of the 771 cases of falsified and substandard medicines brought to the WHO's attention.[29] A 2008 study in Egypt, Lebanon, Jordan, and Saudi Arabia found more than half of antibiotics sampled to be substandard.[30] A similar survey in Burma revealed substandard drugs in 16% of amoxicillin and 13% of ampicillin samples.[31] A recent review estimates that about 35% of the antimalarial medicines in Southeast Asia are substandard, and 36% can be classified as falsified; similar results were obtained in sub-Saharan Africa.[32] The result is growing strains of drug resistant organisms. In Thailand, there is a progressive lengthening of the time it takes for patients to clear malaria parasites from their bloodstream during treatment, suggesting that the parasites are becoming more resistant to artemisinin.[33] Failures have been reported for vaccines and HIV/AIDS medications as well, although for a number of reasons not at similar rates.[34]

The health costs are not borne only by consumers and patients, but by the broad health and legal systems, which are undermined by the criminal networks that exploit global supply chain vulnerabilities. According to the United Nations Office of Development and Crime (UNODC), the sale in falsified medicines supports criminal activities, funds weapons and ammunition acquisitions, and entrenches corruption.[35] UNODC has suggested that in

---

27. Olusola, A. M., Adekoya, A. I., Olanrewaju, O. J., 2012. Comparative evaluation of physicochemical properties of some commercially available brands of metformin HCL tablets in Lagos, Nigeria. Journal of Applied Pharmaceutical Science 2, 41-44.
28. Twagirumukiza, M., Cosijns, A., Pringels, A., Remon, J. P., Vervaet, C., Van Bortel, L., 2009. Influence of tropical climate conditions on the quality of antihypertensive drugs from Rwandan pharmacies. Am. J. Trop. Med. Hyg. Am J Trop Med Hyg 81, 776-781.
29. Wondemagegnehu, E., 1999. Counterfeit and substandard drugs in Myanmar and Viet Nam: Report of a study carried out in cooperation with the governments of Myanmar and Viet Nam. World Health Organization, Geneva.
30. Kyriacos, S., Mroueh, M., Chahine, R. P., Khouzam, O., 2008. Quality of amoxicillin formulations in some Arab countries. Journal of Clinical Pharmacy and Therapeutics 33, 375-379.
31. Wondemagegnehu, E., 1999. Counterfeit and substandard drugs in Myanmar and Viet Nam: Report of a study carried out in cooperation with the governments of Myanmar and Viet Nam. World Health Organization, Geneva.
32. Nayyar, G. M. L., Breman, J. G., Newton, P. N., Herrington, J., 2012. Poor-quality antimalarial drugs in Southeast Asia and sub-Saharan Africa. Lancet Infectious Diseases 12, 488-496.
33. Phyo, A. P., Nkhoma, S., Stepniewska, K., Ashley, E. A., Nair, S., McGready, R., ler Moo, C., Al-Saai, S., Dondorp, A. M., Lwin, K. M., Singhasivanon, P., Day, N. P. J., White, N. J., Anderson, T. J. C., Nosten, F., 2012. Emergence of artemisinin-resistant malaria on the western border of Thailand: A longitudinal study. Lancet 379, 1960-1966.
34. IOM Countering Falsified and Substandard Drugs, 61.
35. United Nations Office on Drugs and Crime, 2009. Transnational trafficking and the rule of law in West Africa: A threat assessment. UNODC, Vienna.

west Africa, the sale in falsified medicines may be worth as much as oil and cocaine trafficking; antimalarials alone accounted for over $400 million in illicit trade.[36]

BARRIERS TO GLOBAL SOLUTIONS FOR THE PROBLEM OF FALSIFIED AND SUBSTANDARD MEDICINES

Despite widespread acknowledgment as to the scope and severity of the falsified and substandard drugs problem, multiple technical and substantive barriers (as well as resource constraints) hinder effective international cooperation on the issue. A principal barrier is agreement about the nature of the problem. For many innovator manufacturers, the problem is primarily one of intellectual property protection. The illicit trade in falsified and substandard medicines (or counterfeit drugs) negatively affects not only pharmaceutical patents but also trademarks and trade dress of drugs. The U.S. Food, Drug and Cosmetic Act, for example, defines a counterfeit drug as a "drug which, or container or labeling of which, without authorization, bears the trademark, trade name, or other identifying mark, imprint, or device, or any likeness thereof, of a manufacturer, processor, packer, or distributor other than the person or persons who in fact manufacture, processed, packed, or distributed such drug and which thereby falsely purports or is represented to be the product, or to have been packed or distributed by, such other drug manufacturer, processor, packer, or distributor."[37] The World Medical Association, by contrast, defines counterfeit drugs as those that are "manufactured below established standards of safety, quality and efficacy and therefore create serious health risks, including death."[38] While both may apply to brand-name and generic products, the bulk ingredients used to make the product, products without the active ingredient, with an insufficient or excessive quantity of the active ingredient, with the wrong active ingredient, or with fake packaging, one decidedly emphasizes the public health aspect of the problem while the other rests on the association between the product and the manufacturer.

The allocation of resources to address "counterfeit" versus "falsified" drugs can vary widely with how these definitions shape debates, legislation, and regulations. For example, in 2012, the United States passed the Food and Drug Safety and Innovation Act, Title VII of which, "Drug Supply Chain," focused on better registration of foreign and domestic facilities, excipient information, interoperable electronic systems, information exchange, standards of admission for imported drugs, registration of commercial importers, notification, risk-based inspections, records for inspection, and processes for

36. IOM Countering Falsified and Substandard Drugs, 73.
37. Federal Food, Drug, and Cosmetic Act. 21 U.S.C.S. § 321(g)(2).
38. World Medical Association, 2012. Counterfeit medical products. Retrieved from http://www.wmanet/en/20activities/30publichealth/50counterfeits.

recognizing foreign government inspections. But the law also imposed new user fees for generics manufacturers that have been perceived as funding FDA enforcement against generics firms related at least in part to the threat they pose to innovator firms. The law expanded the FDA's overseas reach as well as its enforcement powers for foreign facilities. On the topic of an international approach to the problem, the World Health Assembly decided to study the public health effects of counterfeit medicines "excluding trade and intellectual property considerations" precisely because of the divisiveness of the issue at the global level.[39]

Intellectual property concerns aside, there is no standardized vocabulary that reflects the consensus (to the extent that it exists) behind tackling the falsified or substandard drugs problem. European and North American definitions emphasize the closeness with which approved labeling conforms to the so-labeled medical product, whether or not it poses a threat to individual or public health. However, "in many parts of the world, drugs are sold without proper packaging and emphasis on label claims has no practical value."[40] Similarly, the issue of unregistered drugs, while conceptually related to falsified and substandard medicines, actually pertains more to the regulatory capacity of any given national authority and, relatedly, unregistered medicines may be both common and of good quality.

COMBATING FALSIFIED AND SUBSTANDARD MEDICINES THROUGH REGULATORY COOPERATION

In this chapter, we have largely defined the problem as one of falsified and substandard medicines, following the lead of the U.S. Institute of Medicine and concurring with them that "falsified" and "substandard" capture the individual and public health issues at stake without embroiling stakeholders in a heated intellectual property debate. Beyond the vocabulary chosen to address the problem, coordinated, strategic, global approaches using an optimal combination of legal, regulatory, and enforcement measures can help reduce the burden imposed by falsified and substandard medicines.

The first step in combating falsified and substandard medicines is at the prevention level.[41] National regulatory agencies are responsible for overseeing drug safety in their countries and should at the very least maintain a list of all pharmaceutical products authorized for marketing, requiring as part of that process proof of entry into the agency's register including the name and address of

39. World Health Organization, 2010. World Health Assembly Resolution 63(10). Retrieved from *Standard/spurious/falsely-labelled/falsified/counterfeit medical products* available at http://apps.who.int/gb/or/e/e_wha63r1.html.

40. IOM Countering Falsified and Substandard Drugs, 26.

41. Institute of Medicine, Riviere, J. E., Buckley, G. J. (Eds.), 2012. Ensuring safe foods and medical products through stronger regulatory systems abroad. Washington DC National Academies Press.

the manufacturer and/or drug approval applicant and relevant information about the product.[42] The integrity of the supply chain simply cannot be assured if it is not known how products are being brought into the country or by whom. At the national regulatory level, all countries will benefit from adopting common registration materials that may be easily shared and verified. The International Conference on Harmonization Common Technical Document, for example, allows low- and middle-income countries to reduce the costs of applying for the registration of medicines while at the same time providing a mutually recognizable source of product and manufacturer information. Common inspection forms would similarly lighten the burden on both regulators and firms.[43]

National regulatory authorities are also able to undertake at relatively low expense public awareness campaigns about falsified and substandard medicines. The U.S. FDA is undertaking an Internet-based national campaign, BeSafeRx (www.fda.gov/BeSafeRx), to raise awareness about the risks of buying medicines online. In Nigeria, the national drugs authority published lists of known fake products and photos illustrating warning signs in daily newspapers. The campaign is credited with the high rate of public awareness of fake drugs and red flags consumers should look for when purchasing medications.[44]

Aside from unilateral measures that may be taken by national regulatory authorities, there is substantial room for current bilateral, regional, and international coordinating organizations to include low- and middle-income participants or broaden participation where it already exists. Currently, the FDA and European Medicines Agency are undertaking a Mutual Reliance Initiative that not only aims at mutual acceptance of inspection reports, but also hopes to build trust and reliance with personnel based at each agency. A similar partnership is under way between the FDA and Health Canada. The International Coalition of Medical Regulatory Authorities brings together regulators from both low- and high-income countries to discuss mutual problems and raise standards for medicines regulation. The International Conference on Harmonization, which began with the EU, Japan, and the United States over 20 years ago, has forged some important documents that would facilitate medicines registration as well as mutual recognition of inspection information. The Asia Pacific Economic Cooperation organization has developed a supply chain integrity roadmap and has sponsored a related series of workshops on measures against falsified and substandard drugs as well as more general harmonized international standards for the regulation of pharmaceuticals. The Pharmaceutical Inspection Cooperation Scheme gathers inspectorates from 45 countries for purposes of sharing regulatory information and working on joint standards. On a bilateral

42. South African Development Community, 2007. Guidelines for submitting application for registration of medicine. Available at http://medicines.sadc.int/?page_id=119.
43. IOM Countering Falsified and Substandard Drugs, 138.
44. Akunyili, D., 2005. Counterfeit drugs and pharmacovigilance. Paper presented at 10th Pharmacovigilance Training Course held at Uppsala Monitoring Centre, Uppsala, Sweden.

basis, many developed countries have opened overseas offices of agencies like the FDA, opening possibilities for observational and joint inspections for good manufacturing practices, good importer practices, good distribution practices, good pharmacy practices, and pharmacovigilance.

While there are affordable, immediate measures that may be deployed against falsified and substandard medicines, the problem will persist as long as low- and middle-income countries lack the regulatory capacity to implement key measures and technologies. Wealthy countries and their institutions, therefore, must commit to investments in regulatory system strengthening. As FDA Commissioner Margaret Hamburg noted in the foreword to this volume, a strong regulatory system is key for each country's public health, national security, and economic viability in order to promote access to safe, effective, and quality medicine. The IOM detailed one successful public-private collaboration in Tanzania:

> *One promising example of government and private-sector investment in medicine retail is the Accredited Drug Dispensing Outlet (ADDO) program in Tanzania (MSH, 2012). The Tanzanian regulatory authority was eager to discourage illegal stocking and improve dispensing practices at unregistered drug stores through an accreditation program (MSH, 2012; Rutta et al., 2011). The Bill & Melinda Gates Foundation funded the program, which used a combination of training, incentives, and creation of consumer demand to drive changes in the private sector (MSH, 2005; Rutta et al., 2011). Trainers from the Tanzanian Ministry of Health educated drug shopkeepers on proper dispensing techniques, medicines storage, national regulations, business skills, and ethics (MSH, 2005; Rutta et al., 2011). The government offered low-interest loans for improving drug shops, many of which had been stuffy, hot, humid (therefore unsuitable for medicine storage), and not properly secured against theft (MSH, 2012). Participants who met the program's standards were rewarded with legal authority to sell some controlled drugs (MSH, 2005).*

In this chapter, we have implicitly endorsed the U.S. Institute of Medicine's approach to falsified and substandard drugs, including pragmatic calls for direct assistance and for national investment agencies to target support for regulatory capacity in low- and middle-income countries. In a separate report, the IOM has identified gaps and strategies and outlined several core elements of a regulatory system: responsive, outcome-oriented, predictable, risk-based or proportionate, and independent. We believe the IOM's reports represent pragmatic and effectively informed approaches to building regulatory capacity.

CONCLUSION

Falsified and substandard medicines represent the dark side of globalization's push to increase access to medicines for the world's population. The gaps between wealthy and relatively poor countries, particularly regulatory capacity, have redistributed the burden of falsified and substandard medicines onto those who already suffer from inadequate access to health services in general and medicines specifically. In this chapter, we have framed the global problem as one that disproportionately affects low- and middle-income countries and have therefore outlined affordable approaches and called for greater commitments from wealthy countries and donors to invest in developing countries' regulatory capabilities. These are not the only aspects to the problem of course. Expensive and sophisticated technologies like CD3, track-and-trace, holograms, color-shifting ink, watermarks, and RFID identifiers are being developed and deployed successfully in Europe and North America and these technologies represent an important part of solving the false medicines puzzle. Yet these technologies reflect the interests of their developers. Track-and-trace systems rely on the assigning of unique identifying numbers to products. Products that lack identification numbers, or products with identification numbers that cannot be accounted for throughout the distribution chain, must be treated as falsified and removed from the market, even if they come from licensed manufacturers, a measure not only beyond the capability of many low- and middle-income states but also not consistent with the less formal ways medicines are distributed.[45] While acknowledging the usefulness and important role these developments will play, we suggest that given the global scope of the falsified and substandard medicines problem, globalized solutions including investments in regulatory capacity be set at the highest level of priorities.

45. Altunkan, S. M., Yasemin, A., Aykac, I. T., Akpinar, E., 2012. Turkish pharmaceuticals track & trace system. Paper presented at Health Informatics and Bioinformatics (HIBIT), 7th International Symposium.

Chapter 6

Regulatory Capacity in Low- and Middle-Income Countries: Lessons from the H1N1 Influenza Pandemic

Sam F. Halabi,[*,†] John Monahan[‡]

[*]The University of Tulsa College of Law, Tulsa, Oklahoma, USA, [†]Scholar, O'Neill Institute for National and Global Health Law, Washington, District of Columbia, USA, [‡]McCourt School of Public Policy, Georgetown University, Washington, District of Columbia, USA

Chapter Outline

INTRODUCTION

As FDA Commissioner Margaret Hamburg noted in a speech given on July 15, 2014, making smart investments in developing countries' regulatory capacity for food and drugs is critical to both national and international security. Indeed, in a range of episodes highlighted in this volume, investment in low- and middle-income countries regulatory capacity has been advocated as a major solution to enduring global health problems. Strong regulatory authorities in low- and middle-income countries would go far toward thwarting the introduction of false and substandard medicines, ensuring the adequacy of nutritious food supplies, and facilitating access to medicines for both routine treatment

Food and Drug Regulation in an Era of Globalized Markets. http://dx.doi.org/10.1016/B978-0-12-802311-2.00006-0

and public health emergencies. Indeed, it is for the latter type of public health preparedness that we contribute our analysis of the priority that should be given to regulatory capacity in low-resource settings: the global response to the H1N1 influenza pandemic which emerged in 2009 first in Mexico and the United States and then spread quickly around the world.

This chapter focuses on low- and middle-income countries' regulatory capacity and the role it played in the global response to pandemic H1N1 influenza orchestrated by the World Health Organization in partnership with governments and pharmaceutical companies. We address three aspects of low- and middle-income countries' regulatory capacity: (1) the relationship between regulatory barriers and legal barriers that caused pharmaceutical companies to delay participation in the global response; (2) the lack of regulatory capacity that hindered efforts to distribute pandemic vaccine even after legal agreements with pharmaceutical companies and governments were in place; and (3) the role that investment in regulatory capacity in the future could play in establishing timely, harmonized, and efficient procedures for facilitating vaccine and antiviral medical countermeasures consistent with the World Health Organization Pandemic Influenza Preparedness Framework adopted in 2011. We emphasize that investment and planning must occur in noncrisis periods. Understanding the relationships between firms, regulators, governments, and WHO is time-consuming and complex. These relationships and roles must be clear in advance of the next pandemic event.

PANDEMIC INFLUENZA: BACKGROUND AND BASICS OF PUBLIC HEALTH SYSTEM RESPONSE

The historical and global dimensions of influenza-related mortality are astounding and explain why since at least 1918 there have been vocal groups of public health activists who have warned of the catastrophic potential of future influenza pandemics and encouraged investments in community readiness. The 1918 H1N1 pandemic killed between 3% and 5% of the world's population and disproportionately affected healthy adults.[1] The 1968-1969 H3N2 pandemic killed nearly 1 million people and demonstrated the devastating effects (and ease with which) rapid changes in an influenza virus's genetic makeup may level on the global population.[2] Indeed, even within currently circulating influenza strains, including H5N1, there is reason to fear the worst. The 1918 influenza epidemic resulted in an approximately 2% fatality rate.[3] The numbers demonstrate how not only healthcare systems but also entire communities may be quickly overwhelmed.

---

1. Taubenberger, J., Morens, D. 1918 Influenza: the Mother of All Pandemics. Emerging Infectious Diseases 12 available at http://wwwnc.cdc.gov/eid/article/12/1/05-0979_article
2. Cox NJ, Subbarao K., 2000. Global epidemiology of influenza: Past and present. Annu Rev Med. 51:407-421.
3. http://tuoitrenews.vn/society/17722/bird-flu-developments-must-be-updated-hourly-deputy-pm.

Vaccines are the most important line of defense against influenza to prevent infection and control spread of the disease.[4] Before vaccines become available, "community mitigation" measures, such as closing schools and public gathering places, and individual measures, such as wearing masks, buy individuals and communities some time. Unfortunately, many of these alternatives are burdensome on individuals and society and, relative to vaccines and antiviral drugs, offer meager protections against the spread of the disease.

Vaccines are therefore at the core of any effort to mobilize social resources toward the preservation of individual and public health, regardless of whether the disease is affecting populations in rich or poor countries. Yet the process of mobilization in rich and poor countries differs significantly. While wealthy countries may procure pandemic influenza vaccine readily through domestic or nearby manufacturers (and then direct it to well-established networks of distributors and providers), low- and middle-income countries must appeal to countries with vaccine manufacturing capacity or vaccine manufacturers themselves for donated or discounted vaccine doses. Even though recent pandemic influenza threats have originated in middle- or low-income countries, the capacity for pandemic influenza vaccine production is overwhelmingly concentrated in Australia, Europe, Japan, and North America (Figure 6.1).[5]

These regions' pharmaceutical firms are in a persistent cycle of seasonal influenza vaccine production, which is based on surveillance reports detailing which influenza viruses are in circulation, how they are spreading, and how well the previous season's vaccine viruses protect against new strains. When a new influenza strain emerges, the first step in vaccine response is to assess whether the seasonal influenza vaccine will produce adequate immunity to protect against the new strain. CDC researchers concluded that the seasonal vaccine did not protect against pH1N1 in 2009. Pharmaceutical firms, five of which control approximately 80% of the influenza vaccine market, raced to develop a pandemic vaccine.

While there is an effort to diversify and improve vaccine development technology, the process for developing influenza vaccines has not significantly changed for over 50 years. Manufacturers must modify influenza viruses (and for new strains, obtain samples of the virus) to grow efficiently, generally in eggs, so that it may be used for vaccine production. This modification also ensures that the vaccine virus may be safely handled. In order to develop antigens and injectable antiserum to measure vaccine potency, manufacturers must coordinate with reference laboratories and regulatory agencies. Vaccines must then be tested in human trials to assess safety and effectiveness. Regulatory approval for marketing and use is dependent on laboratory-generated evidence and clinical trial outcomes. Even safe and effective vaccines generate adverse

4. Aledort, J., Lurie, N., Wasserman, J., Bozette, S., 2007. Non-pharmaceutical public health interventions for pandemic influenza: An evaluation of the evidence base. BMC Public Health, 7, 208.
5. Crosse, M., 2008. Influenza pandemic: Efforts underway to address constraints on using antivirals and vaccines to forestall a pandemic. *Governmental Accounting Office*, 19. Accessed March 9, 2014 at http://www.gao.gov/new.items/d0892.pdf.

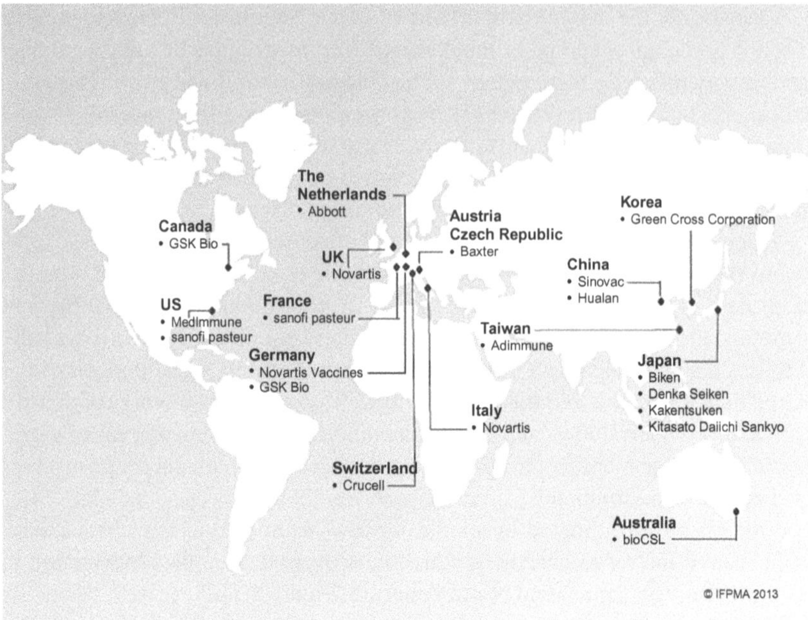

FIGURE 6.1 IFPMA IVS members' production facilities.

events among those inoculated, ranging from (common) soreness at the injection site to fever, discomfort, and muscle pain to (rare) anaphylaxis and oculorespiratory syndrome.[6] Where vaccines are made, and where they may be safely developed, is important. Because manufacturing capacity and associated regulatory oversight are located in only a few countries, the safety profile of vaccines in those regions is quite strong. Where regulatory oversight and experience are inadequate, the safety profile is not as well established or publicized.

PANDEMIC H1N1 AND THE GLOBAL RESPONSE

When researchers in Mexico and the United States concluded that influenza-related hospitalizations in separate, noncontiguous areas of Mexico, southern California, and New York City specially affected children and young adults, they were alerted to the possibility that a new pandemic viral subtype of influenza had emerged (Figure 6.2).[7] After the U.S. Centers for Disease Control and Prevention (CDC) received samples from two early H1N1 patients in mid-April

6. World Health Organization, 2012. Observed rate of vaccination reactions influenza vaccine. Accessed January 28, 2014 at http://www.who.int/vaccine_safety/initiative/tools/Influenza_Vaccine_rates_information_sheet.pdf.

7. Cordova-Villalobos, J.A., Sarti, E., Arzoz-Padres, J., Manuell-Lee, G., Mendez, J.R., 2009. The influenza A(H1N1) epidemic in Mexico lessons learned. Health Res Policy Syst, 7, 21.

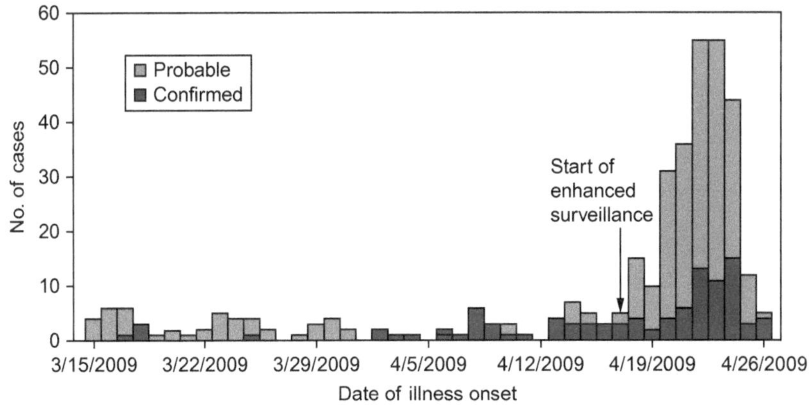

FIGURE 6.2 Number of confirmed and probable cases of swine-origin influenza A (H1N1) virus infection by date of illness onset - Mexico 2009.

2009, researchers exposed banked blood samples taken before and after vaccinations from 2005 to the new virus.[8] Samples from children produced no antibodies, while samples from adults vaccinated against seasonal flu showed a slight increase in antibodies against the H1N1 virus. Because it did not appear that the seasonal vaccine would adequately protect adults against infection, the CDC recommended development of a vaccine specific to the new strain.[9] This recommendation was echoed in WHO's June 11, 2009, declaration of a Phase 6 pandemic (Figure 6.3). Under the WHO classificatory scheme operating in 2009 (it has been revised in light of the H1N1 experience), in Phases 1-3 of a pandemic, influenza circulates predominantly in animals and there are few human infections. In Phase 4, there is sustained human-to-human transmission and in Phases 5 and 6, sustained human transmission spreads to at least two WHO regions.[10]

After the WHO declaration, attention immediately was drawn to the inadequate availability of vaccines in the Southern Hemisphere and other low-income and moderate-income countries that would be unlikely to have resources to purchase vaccines. Because of the timing of the pandemic, pharmaceutical firms were rolling out of their seasonal influenza vaccine cycles and were better positioned to move to pandemic influenza vaccine production. The gene sequence of wild-type pandemic pH1N1 was made publicly available on April 27, 2009. By May 8, 2009, samples of wild-type virus had been sent from reference laboratories to vaccine manufacturers. Vaccines were manufactured all summer, in anticipation of

---

8. Centers for Disease Control and Prevention (CDC), 2009. Serum cross-reactive antibody response to a novel influenza A (H1N1) virus after vaccination with seasonal influenza vaccine. MMWR Weekly, 58(19). Accessed January 28, 2014 at http://www.cdc.gov/mmwr/preview/mmwrhtml/mm5819a1.htm.

9. Hancock, K., Veguilla, V., Lu, X., 2009. Cross-reactive antibody responses to the 2009 pandemic PH1N1 influenza virus. New England Journal of Medicine, 361, 1945-1952.

10. Doshi, P., 2011. The elusive definition of pandemic influenza. Bull World Health Organ, 89, 532-538.

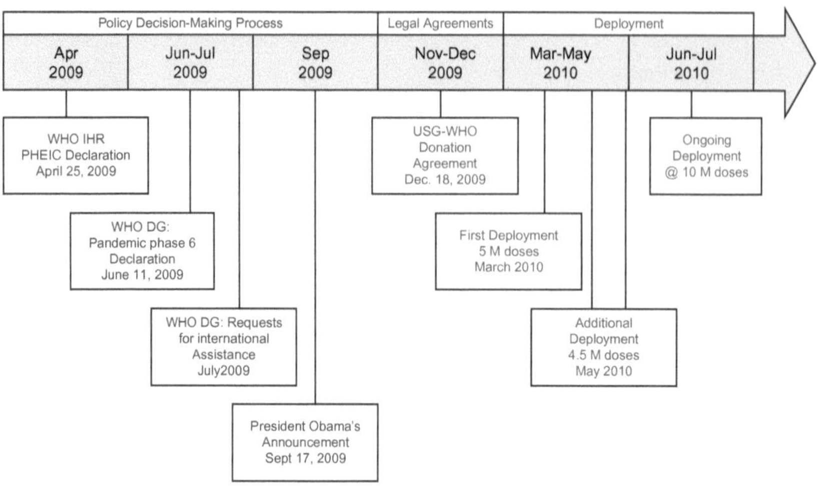

FIGURE 6.3 Timeline for U.S. Donation to WHO in 2009–2010.

H1N1 exacting its most significant human toll during the 2010-2011 flu and cold season in the Northern Hemisphere. Despite the timing advantages, the virus beat the vaccine. Indeed, one of the crucial open questions researchers are attempting answer is what would have happened if a far more virulent influenza strain, unaffected by a seasonal vaccine, hit at a far more disadvantageous time?

REGULATORY BARRIERS AND LEGAL BARRIERS IMPACT PARTICIPATION IN THE GLOBAL RESPONSE

Over the summer of 2009, governments engaged in intense discussions as to how to manage the pandemic internally—for example, the deployment of community mitigation measures and which populations should enjoy priority access to pandemic vaccine—and externally, that is, how countries with substantial manufacturing capacity might facilitate access to pandemic vaccine for those without such capacity. At the U.S. Department of Health and Human Services (HHS), the question as to vaccine distribution or donation to needy countries arose not just in the context of a moral impulse to help with the influenza disease burden, but also because Indonesia had withheld samples of H5N1 avian influenza in 2007 on the basis that it rarely received the benefits (vaccines and antiviral medications) that the WHO's sample-sharing system facilitated for others. On September 17, 2009, President Obama announced the "Plan to Expand Fight against Global H1N1 Pandemic," which committed the United States to provide 10% of its vaccine supply to an international effort.[11]

11. Office of the Press Secretary, 2009. President Announces Plan to Expand Fight Against Global H1N1 Pandemic. *The White House.* Retrieved from http://www.whitehouse.gov/the-press-office/president-announces-plan-expand-fight-against-global-h1n1-pandemic.

The United States also reached out to other countries that wanted to participate in a vaccine sharing system.

At the same time that the United States and its partners were driving a government-side distribution plan, pharmaceutical firms were restructuring their existing arrangements with both governments and the World Health Organization to accommodate the donation, sale, or discounting of pandemic vaccine. The concern that H5N1 avian influenza circulating in Asia might become easily transmissible to, and then between, humans resulted in both divergent (if accelerated) regulatory approval processes as well as a set of agreements entered into between two manufacturers—GlaxoSmithKline (GSK) and Sanofi Pasteur (Sanofi)—and the World Health Organization for donations of prepandemic H5N1 vaccine doses. If the World Health Organization ever declared a Phase 6 pandemic, GSK and Sanofi had pledged 50 million and 60 million doses of H5N1 vaccine, respectively, although no legal agreements for donations were in place. GSK and Sanofi agreed to convert those H5N1 commitments to pandemic influenza A pH1N1 vaccine and to increase the number of doses to 150 million.[12]

The next several months involved discussions of how to produce the vaccine and how to distribute it to the rest of the world. There were three sets of interconnected legal agreements influencing vaccine availability. One set was between WHO and donors, both governments and firms. These presented a variety of liability issues such as: Were manufacturers liable for their product's efficacy? What was their liability relationship with the country making the donation? A second set of legal agreements were required to regulate the relationship between recipient countries and WHO. Each individual country had to enter into negotiations with WHO about conditions of donation, including registration requirements. Finally, manufacturers and governments had to modify existing agreements to comport with WHO's role and the vaccine sharing system generally. All of these negotiations occurred simultaneously.

WHO approached "all known" influenza vaccine manufacturers.[13] Firms found themselves negotiating with WHO over conditions for donation, shipment, and distribution of vaccines. Many governments with preexisting contracts sought to preserve as much of their firms' capacity—that is, firms located within the territorial borders of the procuring governments—as would be necessary to inoculate their populations first before giving or selling to others. As a result, manufacturers negotiated with a much larger than usual number of procurement officials, regulators, healthcare providers, and vaccine distributors.[14] Since at

12. http://www.who.int/immunization/sage/2_Minutes_2nd_SAGE_WG_27-02-11_apr_2011.pdf.

13. World Health Organization, 2011. Implementation of the International Health Regulations. Geneva. Accessed March 9, 2014 at http://www.who.int/ihr/en/.

14. Hanquet, G., Van, D.P., Brasseur, D., De, C., Gregor, S., Holmberg, M., Martin, R., Molnár, Z., Pompa, M.G., Snacken, R., van der Sande, M., Van Ranst, M., Wirtz, A., Neels, P., 2011. Lessons learnt from pandemic A(H1N1) 2009 influenza vaccination. Highlights of a European workshop in Brussels. *Vaccine*, 29,370-377.

least 2006, industry representatives have stated that manufacturers would need advance assurance that governments would provide liability protection in order to donate vaccines. Indeed, some manufacturers reportedly will not even authorize use of vaccine for clinical trials if not insured against legal liabilities. Because the initial urgency of the pandemic response required an unprecedented number of doses of a new vaccine to be deployed globally in a period of only a few months, vaccine manufacturers required that all purchasers or recipients (many of which were European and North American governments) indemnify them for adverse events resulting from the use of the pandemic H1N1 vaccine, with exceptions allowed for failure to follow current good manufacturing processes or other discrete specifications.[15]

From the manufacturers' perspective, these negotiations occurred in the shadow of potentially large liabilities related to their existing contractual arrangements with governments; detailed processes for vaccine approval, distribution, and marketing; as well as more general exposure should quickly developed vaccines generate unexpected adverse reactions or safety problems. Manufacturers were required to seek approval in some jurisdictions as if it were an entirely new vaccine. Because manufacturers faced a range of legal barriers to production, donation, and discounted sale of pandemic vaccines, such as the process by which vaccines may be approved and registered with national regulatory authorities and protection from and indemnification for liability, agreements between pharmaceutical firms, WHO, donor, and beneficiary governments for vaccine sharing took several months. GSK and WHO signed an agreement for the donations on November 10, 2009, which resulted in just over 24 million doses actually donated. Sanofi announced a "flexible" donation of up to 100 million doses on June 17, 2009, but the donation agreement was not signed until December 2009. Novartis specifically eschewed donations, favoring pricing mechanisms to establish a "sustainable way" to deliver vaccines to developing countries.

The role of WHO was fundamentally driven by many countries' lack of regulatory capacity. Because many low- and middle-income countries did not have the ability to screen vaccines or even provide for emergency regulatory procedures, WHO's prequalification system stepped in to fill that gap.[16] Under WHO prequalification, manufacturers submit dossiers including data on ingredients and clinical trial results for certain medicines, and a WHO team assesses the dossiers and inspects products and facilities for compliance with good manufacturing practices. This process exists for a number of reasons. WHO is comprised of member states, and it tries to offer assistance to all of them, especially those

15. World Health Organization, 2010. Main operational lessons learnt from the WHO pandemic influenza A(H1N1) vaccine deployment initiative. Accessed January 28, 2014 at http://www.who.int/influenza_vaccines_plan/resources/h1n1_vaccine_deployment_initiaitve_moll.pdf.
16. Prequalification. *World Health Organization.* Retrieved from http://www.who.int/topics/prequalification/en/.

states that cannot help themselves. Countries without a regulatory agency for medicines and medical devices, or an agency constrained in its technical capacity, rely on WHO. Even medicines and products produced in Europe and North America and regulated by highly competent authorities go through the WHO process, as recipient countries are not conducting inspections; they are relying on the WHO seal of approval.

Even with WHO prequalification available as an important public health service to low- and middle-income countries, relatively few had adopted statutes or regulations clearly authorizing prequalification for medicines generally and for public health emergencies specifically. Only 16 countries rely solely on prequalification while many others include it as a prerequisite. In over half of the beneficiary countries, prequalification of a vaccine by WHO was not sufficient to obtain regulatory approval, and relatively few countries' national laws stated that products donated by the United Nations did not require national registration.[17] So not only did a lack of regulatory capacity impede receipt of donated vaccine, many countries had not even established relatively inexpensive procedures for allowing prequalified medicines to be used in cases of public health emergency. Indeed, two countries did not waive the requirement for *in country* clinical trials and 40 countries requested customized review including lot release certificates, testing levels, and trial data in addition to the certificates confirming good manufacturing practices, legal exportation documents, and certificates of free sale (marketing authorizations) that WHO included as a matter of practice. Because the vaccine was authorized only for pH1N1 and under certain conditions, there was a potential risk of donating or selling vaccine for "off-label" uses, increasing the risk of legal liabilities.

To be sure, not all regulatory obstacles were owed to lack of capacity in low- and middle-income countries. Wealthy governments' advance purchase agreements complicated negotiations between firms and WHO and even label indications in wealthy countries did not explicitly support best practices. With influenza, pregnant women and children are the most at risk and many national regulatory authorities and WHO treat them as priority groups for influenza immunization.[18] Vaccine labels (still) do not reflect these recommendations and, indeed, may subtly discourage immunization of pregnant women by referring to the lack of clinical trial evidence for immunizing pregnant women. For regulators in many low- and middle-income countries less familiar with influenza, the label indication was sufficient to discourage approval particularly in countries where regulators are criminally liable or otherwise accountable for adverse events owed to drugs they approve.

17. WHO, note 14, supra.

18. (2009). Global Alert and Response (GAF): WHO Recommendations on Pandemic (H1N1) 2009 Vaccines. *World Health Organization.* Retrieved from http://www.who.int/csr/disease/swineflu/notes/h1n1_vaccine_20090713/en/.

LACK OF REGULATORY CAPACITY STYMIES EFFORTS TO DISTRIBUTE PANDEMIC VACCINE AFTER LEGAL AGREEMENTS ARE CONCLUDED

Under the deployment regime established by WHO after legal agreements with manufacturers were finalized, a recipient country had to satisfy three prerequisites. First, the country had to submit a letter of intent from the government to WHO officially declaring the country's desire to receive donated vaccine. The second prerequisite was a letter of agreement between the country and WHO. This entailed legal agreement on complex issues that required the government to, among other things, waive liability for donated vaccine. The final prerequisite was that a recipient government had to develop a national deployment plan (NDP) to ensure that the initially scarce supply of vaccines went to countries that were able to effectively use them. The NDP was generally the last of the prerequisites to be completed by countries as few, if any, had a national pandemic influenza preparedness and response plan that included vaccine deployment. As a result, countries required significant time, assistance, and financial resources to finalize an NDP.[19] For many countries, no NDP was ever developed and, indeed, more than 30 countries were unable to provide any information on regulatory approval to WHO at all.

For those countries that did respond, each country's national regulatory authority imposed its own regulatory process for approving H1N1 vaccines, authorizing their importation, and overseeing their distribution. These processes ranged from, on the one hand, one-time waivers of normal rules to, on the other, detailed requirements for pediatric subgroup data, regulatory assessments capacity, quality control preparedness and capacity, and postmarketing safety surveillance and field assessment of efficacy and immunogenicity. Some regulatory agencies approved pandemic vaccines as a type of seasonal influenza vaccine, while others adapted an approval process in place for candidate H5N1 (avian flu) vaccines. The biochemistry of H1N1 vaccines varied, with adjuvanted vaccines (an adjuvant is an inorganic or organic chemical, macromolecule, or entire cell of certain killed bacteria, which enhance the immune response to an antigen) and vaccines produced using cell- rather than egg-based technology facing more significant regulatory review.

Manufacturers required access to information on country regulatory processes that was often difficult to obtain. Reallocating products once this work had begun led to additional work for manufacturers and delayed delivery to countries. In one instance, a change in the delivery schedule necessitated switching to the product of a different manufacturer, which triggered a *de novo* review of all aspects of vaccine approval. In Nigeria, the combination of separate processes

19. http://www.who.int/influenza_vaccines_plan/resources/h1n1_vaccine_deployment_initiaitve_moll.pdf.

for filing vaccine lot release data, obtaining market authorization, and gaining approval for a customs duty exemption delayed receipt of vaccine by 7 months.[20]

The delays caused by this legal wrangling were substantial. For countries in WHO's African Region, vaccines were deployed on average 261 days after a country expressed interest in donated vaccine, a delay attributable to both regulatory obstacles and governmental failures to submit an acceptable deployment plan.[21] "For those countries that were first hit by the emerging pandemic, like those in the Southern Hemisphere, but also for some countries in the Northern Hemisphere, the vaccines clearly came too late and well after the pandemic struck."[22]

Although widely regarded as a process that confronted too many avoidable obstacles for an effective global response to pH1N1, the vaccine sharing system created the conditions that ultimately led to a global agreement on pandemic preparedness. In 2011, the World Health Assembly adopted the Pandemic Influenza Preparedness Framework, which requires not only that pharmaceutical companies jointly fund a vaccine and antiviral sharing system, but also that that they specifically promise to donate real-time vaccine production, facilitate technology transfer, or donate antiviral medications in exchange for drawing from WHO's bank of influenza samples.[23]

THE ROLE OF INVESTMENT IN REGULATORY CAPACITY AND THE EFFECTIVENESS OF THE WHO PANDEMIC INFLUENZA PREPAREDNESS FRAMEWORK

Yet even with a new international vaccine sharing apparatus in place, investment in the regulatory capacity of low- and middle-income countries will be critical to its success and for the global response to the next influenza pandemic. It is not sufficient for manufacturers to promise to donate vaccines if those vaccines cannot be approved, imported, and distributed to those who need them. WHO, as well as North American and European governments, are funding programs to increase the supply of seasonal and pandemic influenza vaccines by expanding global coverage of seasonal flu vaccine, promoting new development sites (including in developing states), and enhancing research and development for novel influenza vaccines.[24] Production capacity is expanding in middle-income

20. World Health Organization, 2010. Main operational lessons learnt from the WHO pandemic influenza A(H1N1) vaccine deployment initiative. Accessed January 28, 2014 at http://www.who.int/influenza_vaccines_plan/resources/h1n1_vaccine_deployment_initiaitve_moll.pdf.

21. (2012). Report of the WHO Pandemic Influenza A(H1N1) Vaccine Deployment Initiative. *World Health Organization.* Retrieved from http://www.who.int/influenza_vaccines_plan/resources/h1n1_deployment_report.pdf.

22. Osterhaus, A.B., Fouchier, R., Rimmelzwaan, G., 2011. Towards universal influenza vaccines. Phil. Trans. R. Soc. B, 366, 2766-2773.

23. http://www.who.int/influenza/pip/PIP_FQA_Nov_2011.pdf.

24. Condon, B., Tapen, S., 2010. The effectiveness of pandemic preparations: Legal lessons from the 2009 influenza epidemic. Fla. J. Int'l. L., 1, 7.

countries like Brazil, Mexico, Thailand, and Eastern European countries, but growing vaccine capacity is not being mapped so that distribution networks can be rationalized nor is there adequate emphasis on countries' regulatory readiness to approve vaccines generally and vaccines for public health emergencies specifically. These preparations should not be taking place in real time during a public health emergency, but should rather be conducted in advance. Even if every capable government and firm wants to donate and every low-income country agrees WHO is the appropriate organization to facilitate access, it still requires substantial resources and planning to operationalize an efficient framework that brings vaccines to countries and populations as quickly as scientific discovery and review would allow.

Universal adoption of WHO prequalification for vaccine in cases of public health emergency facilitates pharmaceutical firms' willingness to participate as do harmonized rules on approval (like clinical trials) and conditions for indemnification or release of liability for adverse events. Even if countries add conditions to prequalification and indemnification or liability release, those conditions should be registered with WHO and available on ministry of health websites so that manufacturers and WHO are not scrambling to meet licensing, marketing approval, and import laws and regulations. Recording such additional requirements also allow countries, manufacturers, and WHO to organize discussions in advance of a public health emergency to address and clarify such regulations to the greatest extent possible. These investments are inexpensive while promising a large payoff in terms of public health readiness. Beyond facilitating vaccine donation, investments in regulatory capacity are critical for low-income countries' plans for vaccine deployment. The time taken for NDPs to be submitted by countries to WHO and for final approval to be granted was no less than 5 months, even for relatively well-resourced countries.

During the pandemic H1N1 episode, each of these aspects of the vaccine development and delivery process generated contracting obstacles when WHO and individual governments approached firms with requests for vaccine donation or purchase. Manufacturers faced differing regulatory and approval processes, uncertain protection from legal liabilities, and equally uncertain and undeveloped systems for distribution even if they could manufacture a limitless number of doses. WHO is optimistic that the agreements put in place between donor governments and firms between November 2009 and March 2010 will provide a time-saving legal framework for production and distribution of vaccines or other medicines during the next pandemic.[25] But there are reasons to doubt that this will be the case based on system-wide response changes. For example, one of the controversial aspects of vaccine development and distribution between 2009 and 2010 was WHO's criteria for identifying a pandemic. Those

25. World Health Organization, 2010. Main operational lessons learnt from the WHO pandemic influenza A(H1N1) vaccine deployment initiative. Accessed January 28, 2014 at http://www.who.int/influenza_vaccines_plan/resources/h1n1_vaccine_deployment_initiaitve_moll.pdf.

criteria were based in some measure on geographic spread rather than severity. WHO has agreed to revise these criteria so that the next time WHO declares a pandemic, it will reflect a more severe public health event on a widespread scale, a scenario likely to render existing legal agreements less applicable than WHO now hopes.

So far as the 2009 pH1N1 experience goes, building capacity without a consistently updated framework for efficiently moving pandemic vaccines from the private sector to the public sphere may simply aggravate the legal and regulatory bottlenecks experienced between 2009 and 2010. The expansion of capacity in middle-income or developing countries enhances the contracting complexity that will likely be faced during the next pandemic. No agreements were reached with firms that are not members of the International Federation of Pharmaceutical Manufacturers Associations (IFPMA), which does not include the small but growing number of manufacturers in developing countries. For the most part, vaccine manufacturers and major purchasers still decide whether to suspend seasonal influenza vaccine production so that all production capacity can be used for pandemic vaccine. Manufacturers also decide whether production of pandemic vaccine can be safely scaled down or suspended in favor of seasonal vaccine. Advance agreements should exist between industry, WHO, and countries as to these decisions or at least create ongoing forums that regularly keep relevant stakeholders current on how vaccine manufacturers' commitments affect overall capacity for production in the case of a pandemic.

Approval processes for national regulatory authorities created a major obstacle not just for initial agreements to donate, but for logistical measures aimed at facilitating deployment of pandemic vaccines as quickly as possible to countries that needed them as soon as possible. As with the vaccine framework developed for H1N1, regulatory harmonization has been largely shaped by pre-2009 preparations for the emergence of a pandemic H5N1 influenza virus strain. WHO, in collaboration with health authorities from Canada, Japan, Spain, and the United States, convened three technical workshops between 2006 and 2007 to examine regulatory harmonization, but the results are shaped by detailed examination of countries with clear regulatory mandates and at least one major vaccine manufacturer. The 2009 H1N1 pandemic has not yet resulted in a measurable increase in agreements between national regulatory authorities or with WHO on data sharing and mutual recognition of some or all aspects of vaccine approval.

CONCLUSION

Regulatory standards and their effective enforcement are essential aspects of health systems. Indeed, it is appropriate and well recognized that countries adopt and implement standards that fit their unique economic, social, and political environments. In the context of a pandemic, whatever the regulatory system, the public health goal is to minimize the time required to apply standards consistent with the nature of the public health emergency. Vaccines are the front line in the global response to the next pandemic influenza outbreak and thus their manufacturers, together with global and national public health authorities, form a critical public-private partnership. The seasonal-pandemic influenza vaccine production balance; the process by which vaccines are developed, researched, and approved for use by regulatory agencies; the potential liability manufacturers face; and the contractual limitations imposed by advance purchase agreements all portend potential delays for the necessary global health response. WHO has already noted that advance agreements between itself, countries, and industries should be negotiated without regard to virus subtype, for a specified period of time (e.g., 3-5 years) and should be regularly reviewed and renewed. Countries that receive donated vaccine, as any purchaser of the vaccine, should adhere to the same practices of releasing and indemnifying manufacturers from certain legal liabilities.

From the foregoing text, we recommend the following principles consistent with the H1N1 experience:

1. Regulatory capacity should be formalized as a priority in private-sector and public-sector driven health aid;
2. Countries that do not have strong regulatory capacity should consider accepting WHO prequalification and formalizing that acceptance in statutes and regulations;
3. For countries that have established and enforced standards beyond WHO prequalification, those standards should be recorded publicly and all efforts should be made to work with vaccine manufactures and WHO to clarify ways in which to expedite application of such supplementary standards;
4. WHO, donor and beneficiary countries, and manufacturers should regularly convene to ensure that interlocking provisions of the legal framework are common, coherent, and well understood by all parties in advance of a major event; and
5. Influenza vaccine labels in all countries should reflect to the greatest extent possible current WHO recommendations as to public health priority immunization groups.

Chapter 7

Prime Minister Narendra Modi's Government and Its Impact on the Indian Pharmaceutical Sector

Vince Suneja*
*TwoFour Insight Group, LLC, McMurray, Pennsylvania, USA

Chapter Outline

INTRODUCTION

In the spring of 2014, the chief minister of the state of Gujarat, Narendra Modi, led the Bharatiya Janata Party (BJP) to an overwhelming victory over the incumbent Congress Party led by Sonia Gandhi in a 6-week (April 7, 2014 to May 12, 2014) general election in India.[1] Modi became the fifteenth prime minister of India on May 26, 2014.[2] Previous to his election, Modi distinguished himself as the leader of Gujarat by implementing policies credited with fostering investment and improving the state's business climate.[3] The national electorate

1. Barry, E., 2014. Narendra Modi's Ambitious Agenda Will Face Difficult Obstacles. The New York Times. Available at http://www.nytimes.com/2014/05/17/world/asia/india-elections.html?_r=0.
2. Gowen, A., 2014. Hindu Nationalist Narendra Modi Sworn in as India's Prime Minister. The Washington Post. Available at http://www.washingtonpost.com/world/hindu-nationalist-narendra-modi-sworn-in-as-indias-prime-minister/2014/05/26/d6f9ba54-25a6-48ac-9693-932132416cf6_story.html.
3. Ahmedabad, 2015. The Gujarat Model How Modi-nomics was Forged in One of India's Most Business-Friendly States. The Economist. Available at http://www.economist.com/news/finance-and-economics/21638147-how-modi-nomics-was-forged-one-indias-most-business-friendly-states.

Food and Drug Regulation in an Era of Globalized Markets. http://dx.doi.org/10.1016/B978-0-12-802311-2.00007-2

viewed his success there as portending similar growth if those policies shaped the national economic agenda. With India suffering from the global economic downturn and the lack of significant progress made by former Prime Minister Manmohan Singh and the Congress Party, the youth and middle class turned out in record numbers to vote for Modi.[4] The business community, especially the multinational pharmaceutical industry that had given up on seeing favorable changes in the Indian pharmaceutical market under Singh and the Congress Party, welcomed the Modi victory.

Notwithstanding Modi's success in the state of Gujarat, the national pharmaceutical sector poses unique challenges related to India's political constituencies and federal regulatory structure. On the one side is the strong export-oriented Indian generic pharmaceutical industry (Indian MNCs) with the aim of becoming the "pharmacy of generic drugs to the world"[5] while on the other is the global multinational innovative pharmaceutical industry (Innovative MNCs) that focus on investing in research and development (R&D) of new drugs. The policy interests of these industry constituencies clash over certain areas like intellectual property rights (IPR) while they align on others like opposition to price controls. There are other competing stakeholders not necessarily aligned with either the Indian MNCs or the Innovative MNCs that the central government must also accommodate when adopting national pharmaceutical policy. Many voters, for example, expect prices for essential medicines to remain affordable, an objective that necessarily implicates India's drugs pricing policies. Other resource and political constraints on the Indian central government make it difficult for any single political party in India to implement wholesale changes to the pharmaceutical sector. In this environment, Prime Minister Modi's trajectory with respect to the pharmaceutical industry will therefore be guided by efforts to balance these domestic interests, as well as demands from international firms and sponsoring governments who seek to participate in India's large market and/or compete with its domestic generics industry.[6] While an analysis of India's pharmaceutical sector under the Modi government might focus on legislative coalitions, global market share growth by Indian generics firms, international trade, and investment diplomacy, or Gujarat-level pharmaceutical policies overseen by Modi during his tenure, this chapter aims to clarify the key ministries and departments that shape the national pharmaceutical sector. Doing so is particularly relevant because, whatever the pressures applied by internal and external constituencies to the Modi government, those pressures will be translated or withstood through India's pharmaceutical regulators who are now

4. Barry, E., 2014. Narendra Modi's Ambitious Agenda Will Face Difficult Obstacles. The New York Times. Available at http://www.nytimes.com/2014/05/17/world/asia/india-elections.html?_r=0.
5. McCambridge, R., 2014. Doctors Without Borders: U.S. May be Trying to Block India's "World Pharmacy." Nonprofit Quarterly. Available at https://nonprofitquarterly.org/policysocial-context/25014-doctors-without-borders-u-s-may-be-trying-to-block-india-s-world-pharmacy.html.
6. Bhalla, N., 2014. Modi Must Resist U.S. Pressure on Drug Patents – MSF. Reuters. Available at http://in.reuters.com/article/2014/09/30/foundation-india-usa-medicines-idINKCN0HP11D20140930.

spread across five key ministries.[7] Indeed, one of the Modi government's initial undertakings was to propose consolidation of many of the regulatory authorities that oversee various aspects of the pharmaceutical approval, marketing, registration, and subsidy agenda.

MINISTRY OF HEALTH & FAMILY WELFARE

The key ministry within the government of India that has the most substantial influence on the pharmaceutical sector is the Ministry of Health & Family Welfare (the Health Ministry). The Health Ministry is a cabinet-level ministry that oversees India's health and family planning policies and is led by Mr. Jagat Prakash Nadda, who was appointed by Prime Minister Modi on November 9, 2014, following a reshuffle of several ministers.[8] The Health Ministry consists of four main departments: Health & Family Welfare; Ayurveda, Yoga and Naturopathy, Unani, Siddha and Homeopathy (AYUSH); Health Research; and HIV/AIDS Control.[9] Before assuming leadership of the Health Ministry, Nadda had served in several posts at the state level in Himachal Pradesh, including as cabinet member for health and family welfare and parliamentary affairs.

The most relevant department with respect to regulation of western pharmaceuticals is the Department of Health & Family Welfare (the Health Department). Under the Health Department sits the Central Drug Standards Control Organization (CDSCO), an Apex Regulatory Authority, led by a drug controller general of India (DCGI).[10] Currently, Dr. G.N. Singh, an accomplished research scientist, holds the drug controller general of India position and oversees the implementation and enforcement of the Drugs & Cosmetics Act, 1940, as amended (the D&C Act).[11] The D&C Act and the associated Drugs & Cosmetics Rules, 1945, as amended (the D&C Rules) (collectively referred to as the Indian Drug Law) provide the framework that governs the regulatory approval mechanism for pharmaceuticals in the country.[12] Under the D&C Act, the regulation of manufacture, sale, and distribution of drugs is largely allocated to state authorities while the national government is responsible for approval of new drugs, clinical trials in the country, establishing standards for

---

7. Other ministries may be relevant over the course of Modi's term. The Ministry of External Affairs, for example, is critical to negotiations with international partners, but for purposes of the topic covered herein, the five ministries named play the most relevant roles.

8. 2014. Jagat Prakash Nadda Replaces Harsh Vardhan as Health Minister. Zee News. Available at http://zeenews.india.com/news/health/health-news/jagat-prakash-nadda-replaces-harsh-vardhan-as-health-minister_1496529.html.

9. Brief History. Ministry of Health and Family Welfare. Available at http://www.mohfw.nic.in/index1.php?lang=1&level=1&sublinkid=1701&lid=1594.

10. Who's Who. Central Drug Standards Control Organization. Available at http://www.mohfw.nic.in/index1.php?lang=1&level=1&sublinkid=1701&lid=1594.

11. *Id.*

12. Acts and Rules. Central Drug Standards Control Organization. Available at http://www.cdsco.nic.in/forms/contentpage1.aspx?lid=1888.

drugs, control over the quality of imported drugs, coordination of the activities of state drug control organizations and providing expert advice with a view of bringing about the uniformity in the enforcement of the D&C Act. Regulation of pharmaceuticals is therefore shared with state-level food and drug administrations.[13] Domestic manufacturing and licenses for marketing authorization are predominantly left to the state FDAs unless the pharmaceutical qualifies as a "new drug," while CDSCO and DCGI have exclusive authority over imports.[14]

The national government's authority over the regulation of clinical trials has drawn both criticism and litigation over perceived deficiencies in oversight and has played an important role in how regulation of the pharmaceutical sector generally is perceived by Indian citizens. In 2009, an HPV vaccination effort undertaken in Gujarat and Andhra Pradesh (and with participation by the states) was accused of obtaining inadequate and irregular consents from vaccinated girls and failing to implement a sufficiently robust adverse event reporting system. A broad range of nongovernmental organizations filed a petition under Article 32 of the Indian Constitution, a so-called public interest litigation (PIL), which allows even parties not injured by illegal conduct to appeal directly to the Supreme Court of India for redress.[15]

Leading up to the announcement of the central government budget on July 10, 2014, the pharmaceutical industry took the opportunity to put forth their wish list of changes, including requesting the prime minister's office to intervene in the ongoing clinical trials court case. The pending PIL has resulted in delays and a significant decrease in the number of trials being approved.[16] During the pendency of the PIL, the Health Ministry through CDSCO and DCGI have introduced a number of changes to the way clinical trials are conducted in India and the obligations of the sponsors, investigators, and ethics committees.[17] Some of the major changes include:

- Mandatory registration of the clinical trial with India's clinical trials registry/database;
- An open-ended requirement for compensation/medical management of injuries occurring during clinical trials, for which an appropriate insurance product may not be available or difficult to obtain;

---

13. The Drugs and Cosmetics Acts and Rules. Ministry of Health and Family Welfare. Available at http://www.cdsco.nic.in/writereaddata/Drugs&CosmeticAct.pdf.
14. 2014. Government Plans to Bring All Pharma-Related Departments under One Roof. The Economic Times.
15. Terwindt, C., 2014. Health Rights Litigation Pushes for Accountability in Clinical Trials in India. Health and Human Rights Journal. Available at http://www.hhrjournal.org/2014/11/06/health-rights-litigation-pushes-for-accountability-in-clinical-trials-in-india/.
16. 2014. India Loses Its Attraction for Pharma Clinical Trials, as Domestic Regulations Tighten. The Pharma Letter. Available at http://www.thepharmaletter.com/article/india-loses-its-attraction-for-pharma-clinical-trials-as-domestic-regulations-tighten.
17. *Id.*

- Unresolved questions related to patient informed consent via audio-visual recording, especially as it relates to privacy concerns and costs associated with obtaining such consent in large trials;
- Review criteria of clinical trials for new drugs:
  - the risk versus benefit of the trial is justified;
  - the trial presents innovation (subject drug) in light of existing therapy options available in the market; and
  - there is an unmet medical need in India that would be served by the proposed clinical trial.
- Creation of a new Apex Committee to oversee approvals of new clinical trials
- Proposal for creation of IT-enabled system for clinical trials.[18]

The uncertainty associated with several of the new requirements has resulted in sponsors and clinical research organizations (CROs) choosing to conduct clinical trials outside of India.[19] The pharmaceutical industry advocated for Modi to use the budget to announce policies and actions that would help end the PIL case at the Supreme Court and streamline the regulatory approval process. Modi declined to do so and, indeed, the budget reversed a service tax exemption for clinical trial related work.

On December 31, 2014, the Health Ministry announced its intention to propose the Drugs & Cosmetics (Amendments) Bill, 2015 (the 2015 Bill) during the upcoming 2015 Spring Parliamentary Session.[20] The 2015 Bill incorporates several of the so-called Office Orders, Notices, and Letters that were issued by the Health Ministry/CDSCO on clinical trials and introduces a definition of medical devices that is separate from "drugs."[21] The latter is important because of the regulation of certain notified medical devices being subject to regulations governing pharmaceuticals. However, the 2015 Bill disappointed the Innovative MNCs as it contained no provisions to centralize the regulatory approval process, implement regulatory data protection (RDP), or create any type of database that might facilitate regulatory findings of therapeutic equivalence and therefore substitutability. The Health Ministry and CDSCO have requested stakeholder comments for both the 2015 Bill and the

18. Saxena, P., Saxena, R., 2014. Clinical Trials: Changing Regulations in India. Indian Journal of Community Medicine 39, 197-202. Available at http://www.ncbi.nlm.nih.gov/pmc/articles/PMC4215498/.

19. Drug Development Blog, 2014. India's Strict Drug Safety Regulations Brought a Decline of Clinical Trials. Aptiv Solutions. Available at http://www.aptivsolutions.com/blog/drug-development-blog/2014/03/indias-strict-drug-safety-regulations-brought-a-decline-of-clinical-trials/.

20. Insights on India—Highlights from the Week of Dec 28, 2014-Jan 2, 2015. TwoFour Insight Group. Available at http://twofourinsight.com/insight-india-highlights-week-dec-28-2014-jan-3-2015/.

21. *Id.*

D&C Rules, providing an additional opportunity to include legislative and regulatory changes.[22]

MINISTRY OF CHEMICALS & FERTILIZERS

The Ministry of Chemicals & Fertilizers (MOCF), which is headed by Ananth Kumar, a long-time BJP politician who served as president of Karnataka state, also plays a critical role in the pharmaceutical industry.[23] Kumar made the November 2014 announcement about potential consolidation of ministries regulating India's pharmaceutical sector. The MOCF is comprised of three departments: the Department of Chemicals and Petrochemicals; the Department of Fertilizers; and the Department of Pharmaceuticals.[24] The Department of Pharmaceuticals (DOP), established in 2008, plays this ministry's most important role with respect to pharmaceuticals as it is charged with formulating and adopting policies that will foster growth of the pharmaceutical sector (both domestically and internationally). Accordingly, the DOP is engaged in policy planning initiatives that take a broad view toward the pharmaceutical sector including foreign direct investment (FDI), education and human resources, research infrastructure, marketing, pricing and access, IPR, and international cooperation.[25] Given the current definition of "drugs" including medical devices, the DOP also claims medical devices as part of their mandate. Over the short tenure of the department, there have been several initiatives launched by the DOP the most important of which are

- Establishment and scale-up of National Institutes of Pharmaceutical Education & Research (NIPER).
- Opening up of unbranded generic drugstores through the "Jan Aushadhi" program (referred to as "generic drugstores").
- Implementation of the Drug Price Control Order, 2013 (DPCO 2013) via the department's attached National Pharmaceutical Pricing Authority (NPPA) office.[26]

NIPERs are public pharmacy research universities established as part of a national investment in training and educating Indians to support its growing and competitive pharmaceutical sector. These institutes offer master's- and

22. Shankar, R., 2015. Health Ministry Releases Draft Drugs and Cosmetics (Amendment) Bill, 2015 to Amend D&C Act, 1940. Pharmabiz.com. Available at http://www.pharmabiz.com/NewsDetails.aspx?aid=86007&sid=1.

23. Who's Who. Ministry of Chemicals and Fertilizers. Available at http://chemicals.nic.in/org5.htm.

24. Ministry of Chemicals and Fertilizers. GOI Web Directory. Available at http://goidirectory.gov.in/ministries_index.php?ct=3.

25. About Us. Department of Pharmaceuticals. Available at http://pharmaceuticals.gov.in/.

26. *Id.*

doctoral-level degrees across the entire pharmaceutical supply chain from medicinal chemistry to business administration degrees focused on bulk purchase and retail. NIPERs also serve as a nexus for collaboration on innovative pharmaceutical research between pharma firms, institute scientists, and affiliated academic centers.

The prescribing, purchasing, and dispensing of drugs in India are typically based on the brand or trade name of the drug versus the international nonproprietary name (INN) or generic drug name. This is predominantly driven by a lack of a central database of all the drugs approved in the country and by whom (e.g., which state FDA), including which drugs are equivalent to each other (e.g., generic version of a brand drug and vice versa). As a result, there is no difference in the marketing of drugs in the country, irrespective of whether the drug is patent protected or not. This set of market circumstances drives pharmaceutical firms to market their drugs to physicians, who have the ability to specify which brand of a drug they prescribe and should be dispensed. In contrast to the United States, where marketing costs for generic drugs are low, generics firms in India must undertake more significant sales and marketing campaigns. The additional cost of marketing, lack of a central repository with information on pharmaceutical bioequivalence, and limited subsidized health insurance programs has resulted in making certain drugs cost-prohibitive, especially for the large number of Indians living in poverty. Several years ago, these factors contributed to the Indian government proposing a generic drugstore concept wherein the majority of the drugs offered to the public would be unbranded, typically less-expensive, generics. The main source of procurement for these unbranded drugs was initially the government-owned pharmaceutical companies known as public sector undertakings (PSUs). The DOP also oversees these PSUs, which have struggled to remain economically viable.

The launch of generic drugstores was met coolly by both industry and the public. The former opposed the generic drugstore concept because it closed some markets to firms while undermining their broader marketing effort. The latter disliked the limited access generic drugstores provided as well as suspecting the stores of stocking lower-quality medicines. Slowly, generic drugstores have started to spread across India and procurement of drugs is no longer exclusively limited to PSUs. Whether these stores will continue to spread across the country and offer patients a quality, less-expensive option (thus influencing the wider market) remains an open question. Parallel to generic drugstores, the national government has also exercised wider authority to control drug prices, including expanding the number of drugs subject to price caps.

In 2013, the DPCO 1995 was expanded to include over 348 essential drugs with a new pricing methodology to calculate the maximum retail price (MRP).[27]

27. Kamath, G., 2014. India's Drug Price Control Regime Just Pulled Off a Logic Defying Move. Quartz India. Available at http://qz.com/270321/indias-drug-price-control-regime-just-pulled-off-a-logic-defying-move/.

Despite discussions taking place with industry almost every year since the original DPCO 1995, industry still was not prepared for the expansion nor the broader policy precedent it set. In early 2014, DOP exercised its authority to bring certain "nonscheduled" drugs such as antiretrovirals (ARVs) under price control.[28] This resulted in litigation being filed by both the generic-based industry association (Indian Pharmaceutical Association, IPA) and the research-based industry association (Organization of Pharmaceutical Producers of India, OPPI) challenging the DOPs decision to expand the list of drugs under price control to include nonscheduled drugs.[29] After the litigation commenced, the DOP withdrew without explanation the internal guidelines that allowed the authority to expand price controls to include certain nonscheduled drugs. While some manufacturers reviewed and altered practices in contemplation of the guidelines, others are waiting until resolution of the lawsuit, resulting in uncertainty in the broader market. While it remains unclear which decisions at which level resulted in the withdrawal of the government's stated position on price controlling nonscheduled drugs, Modi has announced since the withdrawal that it is not the intention of the government to allow prices of medicines to rise.

In addition to discussions on potential expansion of DPCO 2013, the DOP has developed an electronic database, which requires pharmaceutical manufacturers to provide pricing, manufacturing, and production information that is required under DPCO 2013. The DOP also has been active in administrative proceedings and litigation against pharmaceutical manufacturers who overcharged for drugs under price control.

MINISTRY OF SCIENCE & TECHNOLOGY

The Ministry of Science & Technology (Science Ministry), led by Dr. Harsh Vardhan, a surgeon and public health administrator, exercises significant influence over biotechnology-derived products.[30] The Science Ministry oversees three departments: the Department of Biotechnology (DBT); the Department of Scientific & Industrial Research (DSIR); and the Department of Science & Technology (DST).[31] All three departments play a role in the regulation and industrial growth of biotechnology-derived health products, including biologics (both new medicines and so-called biosimilars or follow-on biologics). With respect to biosimilars, DBT was instrumental in soliciting input from various stakeholders to develop India's regulations for biosimilars. The Science Ministry also facilitates incentives for research by awarding grants and providing funding opportunities to

28. 2014. Court Orders Drug Cos to Maintain Status Quo on NPPA July 10 Notification on Price Control, Case Adjourned to Aug 12. Pharmbiz.com. Available at http://www.pharmabiz.com/NewsDetails.aspx?aid=83290&sid=1.

29. *Id.*

30. Department of Science and Technology. Available at http://www.dst.gov.in/.

31. Ministry of Science and Technology. GOI Web Directory. Available at http://goidirectory.gov.in/ministries_index.php?ct=39.

both academic and industry researchers, as well as promoting technology transfer between them. The ministry has also expressed its goal to elevate India's international reputation for scientific discovery and innovation through specific benchmarks by the year 2020. Its most direct impact to the sector is the role it plays through its dialog with the Health Ministry and CDSCO over review of biotechnology medicines. Representatives from the Science Ministry have representative seats on various committees formulated by CDSCO that provide feedback on a range of steps in the pharmaceutical development process from clinical trials to approval requirements for certain biotechnology medicines.

MINISTRY OF COMMERCE & INDUSTRY

The Ministry of Commerce & Industry (MOCI) is led by Ms. Nirmala Sitharaman,[32] a national spokesperson for BJP as well as a former manager at PriceWaterhouse. MOCI, like its equivalents in other countries, is largely charged with promoting economic growth and trade competitiveness, including trade in pharmaceutical products.[33] The ministry divides its responsibilities over pharmaceuticals between two departments: the Department of Commerce (the Commerce Department) and the Department of Industrial Policy & Promotion (DIPP).[34]

The Commerce Department focuses on international trade and assisting industry in exporting products[35] while DIPP focuses on industrial policies, including IPR and other protections relevant to pharmaceutical sector development and interests.[36] DIPP arguably plays a more important role with respect to the pharmaceutical industry as it develops and shapes IPR policies and FDI in the sector. The former has been an issue of concern for the Innovative MNCs in India. Since India's pledge to comply with the World Trade Organization (WTO) Agreement on Trade-Related Aspects of Intellectual Property Rights (the TRIPS Agreement), it has implemented key flexibilities available under the agreement and threatened disputes at the WTO to defend its intellectual property policies that facilitate access to medicines.[37] Innovative MNCs have advocated that the national government

32. Ms. Nirmala Sitharaman is the Minister of State with independent charge for MOCI; as she is not a Member of Parliament, she cannot be a "full" Minister and thus must make do with the relative junior rank of "Minister of State" until such time she is appointed to either the upper or lower house in Parliament.

33. Ministry of Commerce and Industry. Available at http://commerce.nic.in/MOC/index.asp.

34. Ministry of Commerce and Industry. GOI Web Database. Available at http://goidirectory.gov.in/ministries_index.php?ct=6.

35. Organizational Set-Up and Functions. Ministry of Commerce and Industry. Available at http://commerce.gov.in/MOC/about_us_organisational_functions.asp.

36. Role and Functions of the Department of Industrial Policy & Promotion. Department of Industrial Policy and Promotion. Available at http://dipp.gov.in/English/AboutUs/Roles.aspx.

37. 2014. India Pledges New Low-Cost Drugs to Developing Nations. Business Standard. Available at http://www.business-standard.com/article/economy-policy/india-pledges-new-low-cost-drugs-to-developing-nations-114031000198_1.html.

enhance both IP protections and allocate greater resources toward enforcement of those rights including RDP, patent linkage, and additional discretion in issuance of compulsory licenses.

During Modi's visit to the United States in the fall of 2014, the main agenda item for the pharmaceutical industry was the above-noted IPR concerns.[38] Over the course of his visit, Modi agreed to review the IPR framework in India and established the National IPR Think Tank (NITT) with the mandate to formulate India's policy on IPR that takes into account the needs of the country, including economic growth, while ensuring compliance with India's commitments under international treaties and trade agreements.[39] On December 24, 2014, the DIPP revealed the NITT's first draft of the National IPR Policy for India, the content of which disappointed the Innovative MNCs.[40] The NITT does not address or even acknowledge any of the main areas of concern articulated by the industry and instead reemphasizes flexibilities under the TRIPS Agreement (e.g., compulsory licenses) to facilitate access to medicines. It further reiterated its position that its policies were fully compliant with all of its international trade agreement obligations.[41] The draft does recommend specialized IPR courts within India's judicial system, which had been on the list of "asks" for the Innovative MNCs for several years.[42]

MINISTRY OF FINANCE

The current finance minister, Arun Jaitley,[43] who also acts as the minister of corporate affairs[44] and minister of information and broadcasting,[45] heads the Ministry of Finance. A corporate lawyer by training and profession, Finance Minister Jaitley previously represented Cipla in 2008 in a patent infringement case against Roche.[46] Pharmaceutical firms, both domestic and foreign, have taken his appointment as a signal that the Finance Ministry will use its

38. Office of the Press Secretary, 2014. U.S.-India Joint Statement. The White House. Available at http://www.whitehouse.gov/the-press-office/2014/09/30/us-india-joint-statement.
39. *Id.*
40. Subramani, A., 2014. India's Ambitious National IPR Policy Ready; Think Tank Submits Report to Centre. The Times of India. Available at http://timesofindia.indiatimes.com/india/Indias-ambitious-National-IPR-Policy-ready-think-tank-submits-report-to-Centre/articleshow/45653552.cms.
41. Leung, P., 2014. India Releases First Draft of National IP Rights Policy. Managing Intellectual Property. Available at http://www.managingip.com/Article/3414568/India-releases-first-draft-of-National-IP-Rights-Policy.html.
42. *Id.*
43. Finance Minister. Ministry of Finance. Available at http://finmin.nic.in/the_ministry/finance:minister/index.asp.
44. Minister of Corporate Affairs. Ministry of Corporate Affairs. Available at http://www.mca.gov.in/MinistryV2/minister_ca.html.
45. List of Ministers of India. Maps of India. Available at http://www.mapsofindia.com/election/list-of-portfolios.html.
46. 2008. Roche-Cipla Row Test Case for Balancing Health Issues, Patents. First World Pharma. Available at http://www.firstwordpharma.com/node/69565#axzz3POPjjUnz.

significant influence across the government to facilitate the growth of the industry. Pharmaceutical sector players are particularly interested in the implementation of the goods and services tax (GST), raising FDI limitations in the insurance sector (including health insurance), and rolling back the FDI limitations in so-called brownfield pharmaceutical investments/projects. All three of these initiatives appear to be high on the Finance Ministry's immediate priority list.

First, the finance minister during both the monsoon and winter parliamentary sessions expressed the importance of implementing the GST.[47] Because India is a federal republic, direct and indirect taxes are assessed by state and federal governments across the pharmaceutical supply chain. Pharmaceutical supply chain participants organize their product, packaging, and transport contracts so as to maximize tax avoidance strategies, even if doing so requires longer routes or times from order to receipt.[48] The GST would be implemented concurrently by the national and state governments as the Central GST and the State GST respectively. The GST would also bring greater parity to levies on domestic goods and imports. Under a GST, firms could lower the costs and delays imposed by current tax avoidance strategies and rationalize plans for transporting products, building warehouses and distribution centers, and entering into other intermediate agreements.

Second, the majority of Indians are forced to pay out of pocket for pharmaceutical expenses related to their health care. There are a limited number of government insurance schemes at the central and state level that offer insurance coverage for pharmaceutical products. The majority of these programs require patients to procure pharmaceutical products at government hospital pharmacies or limit the coverage to medicines provided in the course of in-patient care. In the private insurance market, the coverage is typically limited to in-patient coverage with a small number of participating hospitals and other providers. With a universal healthcare plan being developed by the central government, there is a promise for a greater market for pharmaceuticals supported by a national insurance or reimbursement scheme. For the pharmaceutical industry, even a minimal universal health care and pharmaceutical coverage plan would spur demand among economic classes now largely excluded from the out-of-pocket payment system. This is especially true for patented pharmaceuticals, which are supported by fairly small market demand and persistently face the threat of compulsory licenses granted by the national government to competitors. Experiences from other states that have rolled out partial or universal healthcare coverage regimes strongly suggest that doing so will result in greater profitability for both innovative and generic pharmaceutical firms.

47. 2014. Arun Jaitley Urges India Inc to Build Public Opinion for GST Bill. Business Today India. Available at http://businesstoday.intoday.in/story/arun-jaitley-gst-bill-ficci-goods-and-services-tax-central-sales-tax/1/213723.html.
48. *Id.*

With respect to FDI in the insurance sector, the Finance Ministry has introduced an ordinance—a type of executive order that requires Parliamentary approval within 6 months—allowing greater foreign ownership in local insurance ventures from the current 26% to 49%, although ownership must remain in Indian hands.[49] Separately, the cabinet during the same session approved a proposal to draw more foreign investment in medical device manufacturers.[50] Under the amended policy, foreign companies may buy a 100% stake in an Indian device maker without government approval, irrespective of whether the investment would be considered greenfield or brownfield. The Finance Ministry's hope is that allowing greater equity stakes in joint ventures will encourage wider provision of health insurance coverage from private-sector providers.[51] Wider private insurance markets would also ease pressure on national drug price controls if joint ventures offer and successfully implement prescription benefit management schemes.

Third, the Finance Ministry has made clear its intent to roll back limitations on FDI in brownfield pharmaceutical projects, which allow acquisitions of equity stakes in existing Indian firms. In late 2011, the Congress Party government curtailed the automatic 100% FDI route for brownfield projects at the urging of the nongovernment organizations (NGOs) in fear of foreign firms closing plants in India, shifting production away from "essential" medicines, increasing the price of pharmaceuticals, and, in essence, taking over the Indian generic pharmaceutical industry that has positioned itself worldwide as a leading exporter of generic pharmaceuticals.[52] Under that regime, FDI in brownfield projects is permitted; however, it requires approval of the Finance Ministry's Foreign Investment Promotion Board (FIPB).[53] That approval is conditioned on certain requirements including prohibiting an investor from imposing competition restraint clauses, as well as idiosyncratic restrictions like commitments to continue the manufacture of specific products.[54] Pharmaceutical companies (both domestic and foreign) have opposed these restrictions as unnecessary and ultimately imposed as a result of unwarranted fears by activist NGOs. Whatever

---

49. 2014. Insurance Bill Likely to Go Through in Winter Session: Arun Jaitley. The Economic Times. Available at http://articles.economictimes.indiatimes.com/2014-11-07/news/55871559_1_finance-minister-arun-jaitley-winter-session-congress-party.

50. 2014. 100% Brownfield FDI will Kill Indian Medical Devices Sector. Millennium Post. Available at http://www.millenniumpost.in/NewsContent.aspx?NID=90489.

51. 2014. FM Arun Jaitley Hopes Insurance Bill will Be Passed in Winter Session of Parliament. Business Today India. Available at http://businesstoday.intoday.in/story/arun-jaitley-on-insurance-bill-in-parliament-winter-session/1/212017.html.

52. Sikarwar, D., 2014. DIPP Wants More Restrictions in Pharma Sector FIPB Nod for FIIs. The Economic Times. Available at http://articles.economictimes.indiatimes.com/2014-06-09/news/50448030_1_dipp-fipb-portfolio-investment.

53. *Id.*

54. 2014. FDI in Brownfield Pharma Stays, More Power to FIPB. The Financial Express. Available at http://archive.financialexpress.com/news/fdi-in-brownfield-pharma-stays-more-power-to-fipb/1217008.

the ultimate legislative and regulatory compromises, the measures announced and advanced by the Finance Ministry signal Modi's and the Finance Ministry's desire to attract new investment in the healthcare sector in India generally and drugs and medical devices specifically. The growing presence of international partners, as foreign direct investors or otherwise, will inevitably influence the evolution of the Indian drug laws. Those laws will increasingly be influenced by policies that balance equity investments with Indian management control as well as what appears to be the position of the Modi government that increased private-sector provision of products and services may at least partially and beneficially displace central government regulatory or subsidy involvement.

EMERGING PATTERNS IN THE PHARMACEUTICAL REGULATORY ENVIRONMENT UNDER THE MODI GOVERNMENT

Early moves by ministers and subordinate directors at Health, Chemicals & Fertilizers, Commerce, and Finance strongly suggest that an overhaul of the pharmaceutical regulatory system is a key item on the Modi government's agenda, with several early initiatives focused on the pharmaceutical sector including consolidation, facilitation of foreign investment, and supplementary insurance schemes to support direct government control. Whether these initiatives can muster sufficient support in Parliament to mature into a comprehensive national pharmaceutical strategy remains an impossible outcome to predict. Legislators are ultimately accountable to their constituents and the changes at the national level necessarily implicate traditional state authority over (and revenue from) drug registration and marketing along with the impact of central policies on small- to medium-size pharmaceutical companies in each state. This makes garnering enough support in Parliament for the changes announced much more difficult. In addition, the question remains as to whether the changes already announced are enough of a sign to foreign pharmaceutical companies that India is serious about investment in the pharmaceutical sector and that the domestic market will grow as projected. Even against the backdrop of these looming questions, however, it is clear that the Modi government intends on giving the pharmaceutical sector a privileged place in its policy deliberations.

Chapter 8

Regulatory Challenges for Medical Devices and Drugs Under the China Food and Drug Administration

Gordon B. Schatz
Healthcare, China Trade Strategies, Washington, District of Columbia, USA

Chapter Outline

On November 10, 2014, U.S. President Barack Obama and China's President Xi Jinping announced an agreement to lower tariffs on various high-technology products, including the 8% tariff on magnetic resonance imaging (MRI) systems.[1] The length of the negotiation, specificity of the products subject to the agreement, and respective domestic constituencies represented in the deal, demonstrate the complex financial, regulatory, and economic

1. Davis, B., 2010. U.S. China to Drop Tariffs on Range of Tech Products. Wall Street Journal (These actions at the 2014 Asia-Pacific Economic Cooperation Summit in Beijing require World Trade Organization ratification).

Food and Drug Regulation in an Era of Globalized Markets. http://dx.doi.org/10.1016/B978-0-12-802311-2.00008-4

challenges facing pharmaceutical and medical device firms aiming to commercialize innovative products in China (Figure 8.1). Because China's market for medical devices and drugs is one of the fastest growing in the world, clearing regulatory, business, and financial hurdles is crucial for product success. Understanding the China Food and Drug Administration (CFDA) is a priority regulatory agency for multinational drug and device manufacturers who want to sell to Chinese hospitals, doctors, and consumers. In an era of globalized markets, careful engagement with the CFDA is a critical starting point for market access (Figure 8.2).

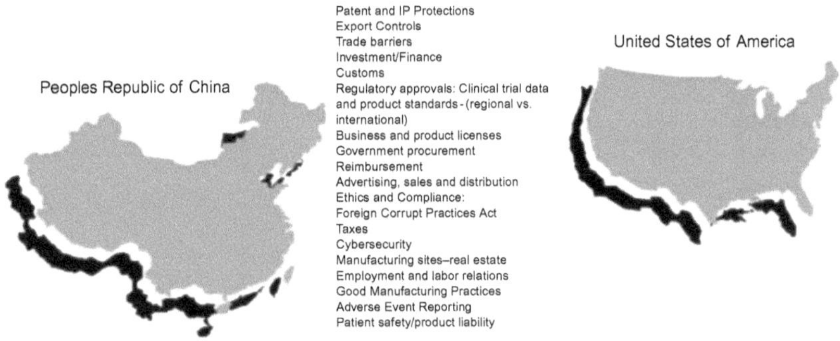

FIGURE 8.1 Legal and regulatory issues in a global market—China and the United States.

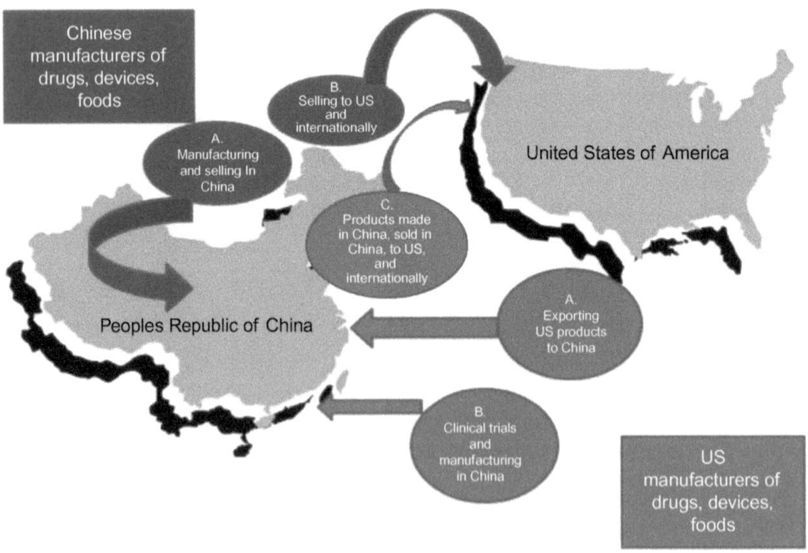

FIGURE 8.2 Global business and regulatory flow—China and US.

CHINA'S POLITICAL AND REGULATORY GOALS: THE CONTEXT FOR DRUG AND DEVICE REGULATION

Businesses entering the China market must understand the Chinese government's political and regulatory goals. China seeks to transform its economy from one historically based on lower-cost manufacturing of consumer goods for global markets (e.g., textiles and clothes), to higher-technology product development accompanied by more sophisticated manufacturing.[2] Paradoxically, China is also committed to full employment, economic stability, and growth; commitments better supported by traditional labor-intensive industries. China's policies seek to balance economic diversification with social stability and harmony. For example, manufacturing smartphones and related technology fits within China's growth trajectory as long as it is supported primarily by manual assembly; transitioning to robotic processes may displace workers.[3] State support for manufacturing operations, therefore, may be implicitly or explicitly conditioned on compliance with background growth objectives. Similarly, because China desires to expand its role as a major exporter of high-technology products to the world, CFDA regulations may subtly favor local Chinese companies over multinational companies.

China's goals in the healthcare sector are to expand the availability of health services from its first- and second-tier cities (Shanghai, Shenzhen, Guangzhou, Beijing, Dalian, Tsingtao), in the highly urbanized eastern coastal regions, to the central and western rural parts of China. These goals are similarly shaped by China's wider strategy for development, particularly the aim to stimulate domestic demand. Chinese residents in the interior who lack access to basic medical services save to cover expenses in case of catastrophic illness instead of purchasing increasingly available consumer goods.[4] There is thus not as high a priority on advanced technology medical products, as on basic health services for 1.3 billion Chinese citizens.

This chapter presents an overview of the most important CFDA regulations and suggests practical tips for managing regulatory hurdles, endeavoring to place those regulations in the appropriate economic, political, and social context. Embedded within these key regulations are strands of the national economic policy and thus cues as to where businesses are most likely to comply and therefore participate in the Chinese market at relatively lower cost. These cues are made explicit in the form of "practical tips" at

2. *See*, United States-China Business Council, 2013. China's Strategic Emerging Industries: Policy Implementation, Challenges, and Recommendations. Uschina.org (Biotechnology is one of the seven strategic emerging industries).

3. Dowd, R., 2014. Apple's iPhone 6, made in part for Apple in China by Foxconn's electronic robots. The Fiscal Times.

4. Christina Ho, "Health Care Reform and De Facto Federalism in China," 8 China: An International Journal 33 (2010).

the conclusion of the chapter. While this volume is primarily committed to pharmaceutical regulation, related and similar regimes for medical devices are included to provide an additional regulatory area where the economic and political context of regulation may be ascertained. Therefore, the chapter begins with pharmaceutical regulation then provides an assessment of medical device regulation.

DRUGS

Patented and Generic Drugs

A "drug" is defined as an article that is used in the prevention, treatment, or diagnosis of disease in humans and that is intended for the regulation of the physiological functions of humans.[5] For drug registration/approval purposes, drugs are classified into three categories:

- Chemical drugs
- Biological drugs (including therapeutic and preventive drugs)
- Traditional Chinese medicines (TCMs) and natural drugs

All new or generic drugs must be registered and approved by the CFDA before they can be imported, manufactured, and advertised in China. The CFDA's Provisions of Drug Registration set different registration and approval procedures for new drugs and generic drugs. New drugs in China are defined as "drugs that are never sold on the China market." New drugs may or may not be protected under a patent in China. The most significant difference between the registration and approval procedures for new drugs and generic drugs lies in clinical trial requirements: new drug registration and approval generally require completion of all phases of clinical trials while generic drug registration and approval require less-extensive clinical trials.

The application for a New Drug Certificate or a Drug Manufacture Permit requires patent information, formula, manufacturing processes, and uses. If another party owns the patent in China, the applicant provides a statement of noninfringement.

The applicant for registration of a generic drug must be a drug manufacturer, and the drug must be within the scope of the Drug Manufacture Permit. The generic drug must have the identical active ingredients, route of administration, dosage form, strength, and therapeutic effects as the registered new drug. The CFDA may suspend the review and approval of a generic drug if the marketed drug has safety problems.

---

5. *See* Drug Administration Law of the People's Republic of China (No. 45), Article 102. China Food and Drug Administration. Available at http://eng.sfda.gov.cn/cmsweb/webportal/W45649037/A48335975.html. *See also* Provisions for Drug Registration, SFDA Order No. 28. China Food and Drug Administration. Available at http://eng.sfda.gov.cn/WS03/CL0755.

Registration

The approval for drug manufacture issued by the CFDA is valid for 5 years and if the applicant intends to continue manufacturing the drug it must apply for re-registration within 6 months prior to expiration. Compared with the initial registration/approval, the re-registration generally requires less data and may not require clinical trials.

The CFDA is recognizing expedited procedures for innovative drugs through fast-track registration and approval. See Provisions for Drug Registration, Chapter IV, and http://eng.sfda.gov.cn/WS03/CL0757/78704.html (February 28, 2013). The CFDA will consider the following:

- Active ingredients extracted from plants, animals, or minerals that are newly discovered medicinal materials and that have not yet been marketed in China;
- Chemically active pharmaceutical ingredients and biologics that have not been approved for sale either in China or outside China;
- New drugs for treatment of HIV, malignant tumors, and rare diseases that have significant clinical advantages;
- New drugs to treat those diseases that do not have effective alternate therapies.

China has adopted a licensing system for drug manufacturers, distributors, and medical institutions. See the English language version of the Drug Administration Law (No. 45, December 1, 2001) and the Regulations for Implementation of the Drug Administration Law, Chapters II, III, and IV (effective September 15, 2002) available at http://eng.sfda.gov.cn/WS03/CL0767/61640.html.

A drug manufacturer must obtain a Drug Manufacture Permit granted by the local Food and Drug Administration (local FDA) before it registers with the local Administration of Industry and Commerce (local AIC) and starts to manufacture its drug. The Drug Manufacture Permit sets forth the scope of manufacturing, is valid for 5 years, and can be renewed. The drug manufacturer must have

- legally qualified pharmaceutical and engineering professionals and the necessary technical workers;
- the premises, facilities, and hygienic environment required for drug manufacturing;
- the facilities, instruments, equipment, and personnel capable of quality control, manufacturing, and testing for drugs; and
- standards and procedures to ensure the quality of its drugs.
- The drug manufacturer shall comply with the Good Manufacturing Practice for Drugs (GMP).

Similarly, a drug distributor cannot register with the local AIC without a Drug Distribution Permit granted by the local FDA. The Drug Distribution Permit is also valid for 5 years, and shall be renewed upon expiration.

Drug distributors must comply with China's Good Supply Practices (GSP).

Clinical Trials

A new drug developer must conduct clinical trials for the new drug in order for the new drug to be registered and approved by the CFDA.[6] The clinical trials consist of phases I, II, III, and the after-market phase IV.

Clinical trials must be conducted at qualified clinical trial institutions (primarily hospitals certified as such qualified institutions). Drug clinical trials, including bioequivalence study, are also subject to the CFDA's approval. The dossier on the new drug research and development including, among other materials, the study methodologies, quality parameters, results of pharmacological and toxicological study, and the related data and the samples shall be submitted to the CFDA for the approval. The clinical trials include phases I, II, and III tests and the after-market phase IV test. The application and approval procedure for drug clinical trials is shown in Figure 8.3.

The drug developer (i.e., clinical trial sponsor) shall conduct the clinical trial within 3 years after the CFDA's approval. The clinical trial must be conducted at qualified clinical trial institutions (primarily, hospitals in China) published by the CFDA, and in compliance with the Good Clinical Practice. The clinical trial drug used in the clinical trial shall be provided by the sponsor

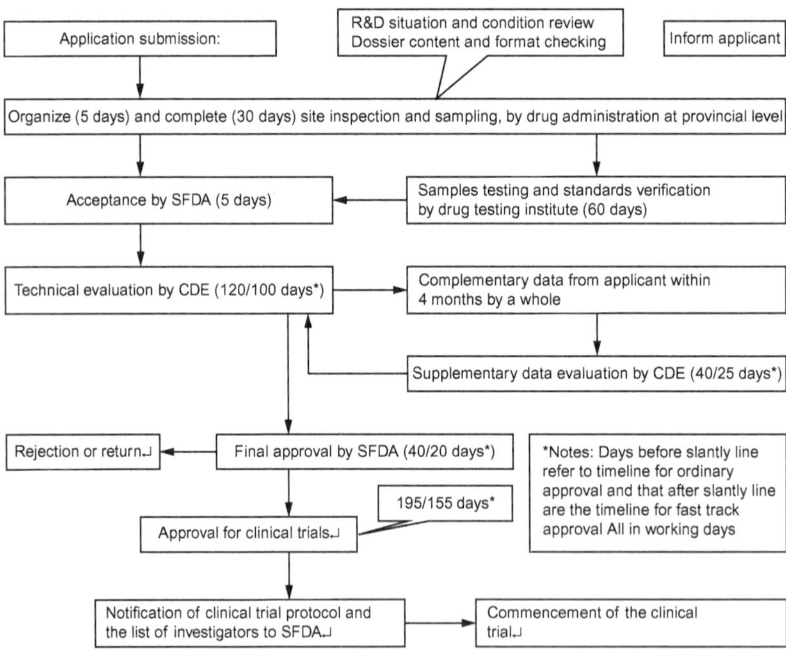

FIGURE 8.3 Application and approval procedure for drug clinical trials.

6. Provisions for Drug Registration, Section 2.1. China Food and Drug Administration. English language version available at http://eng.sfda.gov.cn/WS03/CL0768/61645.html.

and must not be sold. The clinical trial drug must be manufactured at a GMP-compliant facility.

Manufacture and Sale of Drugs

The CFDA and its local counterparts are responsible for drug regulation in their administrative areas, which includes registration of drugs, approval for drug manufacture, approval for drug advertisements, supervision of drug sales, granting Drug Manufacture Permits to companies for manufacture of drugs, Drug Distribution Permits for wholesale and/or retail distributors, GMP certification and GSP certification, and so on.

Local AIC agencies also regulate many features in the development and commercialization of drugs.

The National Development and Reform Commission and its local counterparts regulate pricing of drugs covered by medical insurance. The National Health and Family Planning Commission and its local counterparts oversee drug procurement by public hospitals.

Drugs That Have Been Licensed/Approved in Other Jurisdictions

Similar to drugs that have not been licensed/approved in any jurisdiction, drugs that have been licensed/approved in other jurisdictions must also complete the registration and/or approval with the CFDA before they can be manufactured or sold in China. Companies intending to manufacture or distribute such drugs must also have a Drug Manufacture Permit or Drug Distribution Permit. However, the clinical trial requirement for registration and/or approval for drugs that have been licensed/approved in other jurisdictions could be less extensive.

Drugs manufactured in China are generally permitted to be exported to other jurisdictions without special approval or license. Radioactive drugs, narcotic drugs, psychotropic drugs, and drugs that fall within the catalog published by the CFDA from time to time are, however, subject to the issuance of export approval or drug sale certificate by the CFDA or its local counterparts.

Drugs manufactured in other jurisdictions can be imported into China, provided that the drug to be imported must first be registered with the CFDA and obtain the Import Drug Registration Certificate for the drug. The applicant for the Import Drug Registration Certificate is the manufacturer of the drug. Similar to the registration and approval procedures for new drugs and generic drugs manufactured in China, the import drug registration also involves submission of required materials as well as completion of clinical trials within China. The Import Drug Registration Certificate is valid for 5 years and the certificate holder must apply for re-registration with the CFDA within 6 months prior to expiration of the Import Drug Registration Certificate. The clinical trial requirements vary, depending on whether the drug to be imported has been approved for sale in the country of origin, whether China has a national standard for the drug to be imported, and so on. In addition to the Import Drug Registration Certificate, import approval is also required for import of drugs under special regulation such as radioactive drugs, narcotic drugs, and psychotropic drugs.

Companies engaged in drug export or import must hold a Drug Manufacture Permit or Drug Distribution Permit. Imported drugs are required to enter into China through designated ports and go through other import formalities such as inspection and customs clearance.

Advertising

Prescription drugs may only be advertised in medical or pharmaceutical professional publications that have been designated by the CFDA and the National Health and Family Planning Commission. Direct-to-consumer advertising of prescription drugs, however, is prohibited. Over-the-counter (OTC) drugs can be advertised to consumers, but only with advance government approval. Advertising is prohibited for radioactive drugs, narcotic drugs, psychotropic drugs, toxic drugs for clinic use, preparations made by medical institutions (i.e., hospitals), drugs for military needs, and drugs manufactured on a trial basis.

Approvals are required and content is regulated. The manufacturer or distributor must obtain an approval and an approval number for drug advertisements from the local provincial-level FDA where the domestic drug manufacturer is located (for a domestic drug) or where the drug import agency is located (for an imported drug). The content of the drug advertisement must be truthful, and comply with applicable CFDA Drug Administration Law requirements, and also China's more general Advertising Law (1994). See also, Provisions for Drug Advertisement Examination (SFDA Decree No. 27, May 1, 2007). The content of the advertisement cannot present indications that exceed the CFDA-approved package insert. For prescription drugs, the advertisement shall state that "this advertisement is only for medical or pharmaceutical professionals," while for OTC drugs, the advertisement shall state "please purchase and use in accordance with the drug instructions or under the guidance of a pharmacist" and also include the symbol for OTC.

- The advertisement must not exaggerate the drug's therapeutic effectiveness or seriously mislead or cheat the customer;
- The advertisement cannot compare itself with other drugs;
- Drug advertisements shall not be published in publications for minors, broadcasting, TV channels;
- Drug advertisements must not contain comments/endorsements from medical institutions, or offer free clinical services, or clinical hotline consultations.

If the advertisement is to be published in a province different from where the domestic drug manufacturer or the drug import agency is located, a record shall be filed in advance with the local FDA of that province. If the applicant for a drug advertisement is a drug distributor, it shall obtain the consent of the manufacturer.

Once an advertisement is approved, there can be no change in the contents of the advertisement, unless the company obtains an approval.

There is no clear distinction between advertising and nonadvertising promotion. The general requirements under the Consumer Protection Law will apply to nonadvertising promotion and generally require that the information not be false or misleading.

Anticorruption Compliance

In 2013 and 2014, there have been heighted investigations and regulatory enforcement actions against multinational drug companies on allegations of antibribery and anticorruption activities. In 2014, for example, China imposed a $490 million fine on GlaxoSmithKline for alleged payments to physicians and hospitals to purchase GSK drugs, funneled through travel agencies.[7] In this case, a court sentenced five GSK China managers to 4 years in prison. These sentences, however, were subsequently suspended.

Not only must manufacturers be alert to compliance with Chinese anticorruption laws, but also U.S. Foreign Corrupt Practices Act restrictions that can arise out of the same marketing and sales practices. The FCPA prohibits many financial dealings between a U.S. company and any Chinese government employee. This can create significant exposure as most Chinese hospitals are government owned and physicians are typically government employees. In a country where business courtesies are a long-standing cultural feature of business dealings, companies conducting clinical trials in China, working with key opinion leaders, and dealing with CFDA officials must be alert to entertainment, meals, gift-giving, support for professional conferences, and retaining consultants to ensure compliance with Chinese and U.S. laws.

DEVICES

The regulation of devices is governed primarily by the CFDA and implemented at local, provincial, or national levels.[8] CFDA regulations recognize three classes of medical devices: Class I, Class II, and Class III, based on progressively

7. Bradsher, K., Buckley, C., 2014. China Fines GlaxoSmithKline Nearly $500 million in Bribery Case. New York Times.

8. CFDA also oversees clinical trials, manufacturing, and post market surveillance. Other Chinese agencies have exclusive or shared responsibilities for importing, payment, sales controls, and corporate operations, including the National Health and Family Planning Commission, State Administration for Industry and Commerce, and General Administration of Quality Supervision, Inspection and Quarantine (GAQSIQ).

increasing health risks. Class I devices are those for which safety and effectiveness can be ensured through routine administration, like disinfecting devices; Class II devices require further controls to ensure their safety and effectiveness, for example weak laser *in vitro* treatment instruments; Class III devices are implanted into the human body, or used for life support or sustenance, or pose potential risk to the human body and thus must be strictly controlled for safety and effectiveness. Any form of implant, for example, would typically be considered a Class III device. Class III and imported medical devices must be registered/approved by the CFDA at the national level. Class I and II devices may be subject to regulation by the provincial or local CFDA.

Importing and Registering Medical Devices

Medical devices cannot be imported into China without first receiving a CFDA registration certificate, which may require type testing in China. To import a medical device into China for CFDA registration-type testing, the agent for the device should submit the instructions for use, quality standards, testing methods, product samples, and marketing authorization certificates to the CFDA testing center, signing a testing contract with the center. The testing center will issue a temporary Import Product Registration Certificate to the agent, which can be used to apply for customs approval.[9]

Additionally, a manufacturer must determine the product registration classification. The procedures for initial registration vary depending upon whether the device is approved or not approved outside China and whether the device qualifies as a Class I, II, or III medical device.[10] Class I and some Class II devices may be exempt from registration-type testing if they have been approved by the governmental medical device authority at the place of their origin, and if the product meets the following[11]:

- The basic theory, main functions and structure, materials used, and anticipated purpose of the medical devices under application for registration and those of its medical devices approved for registration fall in the same class;
- It has passed the inspection on its quality management rules for production of medical devices or has obtained the quality system certification for

9. Regulations for Supervision and Administration of Medical Devices, Article 11. China Food and Drug Administration. Available at http://eng.sfda.gov.cn/WS03/CL0767/61641.html. *See also* Initial Registration of Import Products. China Food and Drug Administration. Available at http://eng.sfda.gov.cn/WS03/CL0770/61661.html.

10. Note that CFDA issued a number of new regulations on December 5, 2013, including Initial Registration for overseas medical devices that are approved for marketing abroad, Initial Registration for overseas Class I devices not approved for marketing abroad and Initial Registration for overseas Class II and III medical devices. Available at http://eng.sfda.gov.cn/WS/03/CL0770/98145, 98142 and 98138.

11. Measures for the Administration of Medical Device Registration, Article 13. China Food and Drug Administration. Chinese language version available at http://www.sfda.gov.cn/WS01/CL0053/25844.html.

medical devices, and is able to provide the testing report accredited by the former institution for examination of enterprise production conditions;
- When the medical devices under application for registration are compared with its products of the same class that have been approved for registration and have passed a registration test, there is no change in its safety or effectiveness, or although if there is change in the safety or effectiveness, both the changed part and the part causing other relevant changes in the safety or effectiveness of the products have passed the test carried out by the medical device testing institution;
- No serious adverse event is found from the adverse event monitoring conducted as prescribed upon its products of the same class that have been approved for registration;
- No records of being unqualified are kept on its products of the same class that have been approved for registration within 1 year from the selective supervisory product quality inspection made by the food and drug administration.

Clinical Trials

Class II and Class III medical devices generally require safety and effectiveness clinical trial data from at least two CFDA-designated clinical trial sites. Devices imported into China require clinical trial data, which in some cases can be from clinical trials conducted outside of China.[12] Clinical trials of medical devices are divided into:

- Clinical testing—new device principles and performance can guarantee safe and effective operation;
- Clinical verification—verifying that the product is virtually the same as other products in the market in terms of structure, performance, safety, and effectiveness.

Clinical trials require that the product meet or have the following:

- Applicable registered product standards or corresponding state or industrial standards;
- A self-examination report;
- A qualified testing report from the testing center recognized by CFDA or GAQSIQ;
- Animal testing reports if the product is to be implanted into the human body for the first time;
- If animal experiments are needed to verify the safety to the human body, the animal testing report must be submitted.

12. Clinical Trials of Medical Devices, SFDA Order No. 5. China Food and Drug Administration. Available at http://eng.sfda.gov.cn/WS03/CL0768/61644.html.

The clinical trials must adhere to the following:

- Implementer (manufacturer) to work with the medical institutions to prepare the protocol. Data for CFDA can only be accepted from government-designated research medical institutions.
- Provide products for free
- Train personnel
- Obtain informed consent from patients, notify patients of the progress of the trial, and comply with Helsinki Declaration
- Report adverse events
- Compensate test subjects for any injuries

Advertising Medical Devices

To advertise a medical device, the medical device manufacturer or distributor must first obtain an approval and an approval number for the advertisement from the local or national FDA before issuing the medical device advertisement.[13] The local Administration for Industry and Commerce (AIC) also has the authority to take enforcement actions against illegal advertising. The content of the advertisement must conform to the CFDA-approved indications for use, and otherwise be truthful. The approval number is valid for 1 year.

Medical device advertising must be scientific and accurate, and cannot:

- Assert or guarantee curative effects, effectiveness ratios or cure rates;
- Compare with the functions and safety of any other medical apparatus, drug, or therapy;
- For personal use devices, use terms that could take advantage of consumers' lack of professional and technical knowledge;
- Contain such information as "it is discovered upon research" or "the experiment or data proves";
- Indicate that the product may cure all diseases and is applicable to all symptoms;
- Contain such promises as "safe," "no toxic or side effects," "refund guaranteed," "no dependence" or "guaranteed by insurance company," or containing such absolute or exclusive expressions as "only," "precise," "the most advanced technology," "the most advanced science," "national-level product," and "never seen in the domestic market";
- Indicate that the product is a necessity for living a normal life or curing diseases;
- Indicate that using the device may help people cope with modern living pressures or academic pressure and will improve their test scores, make them full of energy, increase their competitiveness, or make people taller and more intelligent;
- Contain clinical service information offers of free service, or implied endorsements by using a medical institution's or doctor's name;

13. *See* Ministry of Health of the People's Republic of China, State Administration for Industry and Commerce of the People's Republic of China, China Food and Drug Administration, 2009. Criterion for the Examination and Publication of Medical Apparatus Advertisements.

- Target children or be published in publications or media aimed at minors;
- Contain "please read the product instructions carefully" or "please purchase and use under the guidance of medical personnel" for personal use devices;
- Contain "see instructions for contradictions or cautions," if on the CFDA labeling.

Violations of the advertising regulations result in a fine not exceeding 10,000 yuan or in some cases up to 30,000 yuan.

New Device Regulations—State Council Order No. 650

Starting June 1, 2014, China's regulation of medical devices advanced with the new Administrative Regulations on the Supervision and Administration of Medical Devices (State Council Order No. 650, released March 7, 2014).[14]

The new regulations[15] provide for

- More streamlined filing process for Class I devices.
- Updates in clinical trial requirements with the possible future issuance of a catalog identifying devices not subject to clinical trials.
- Potentially more expeditious sequence of getting a medical device registered first, then getting the manufacturing license.
- Compliance with GMPs and the implementation of quality control systems now include self-inspections and submission of a manufacturer's self-inspection report to the CFDA's counterparts at the provincial level.
- Significantly higher penalties for violation of requirements.

Registration and Filing Requirements

There is a more expedited filing procedure for lower-risk Class I medical devices, including those manufactured in foreign countries. Higher-risk devices in Class II and Class III still require registration, including submission of comprehensive product data and type testing.

Under the old regulations, a material change in a registered device required a new registration as a new device.[16] By comparison, under the new regulations, a material change only requires a change in registration. Generally, a material change in a registered medical device is when the device's safety and/or efficacy

14. *See* China Food and Drug Administration. Available at http://eng.sfda.gov.cn/WS03/CL0755 (The CFDA references below are to the English translations of selected CFDA regulations and notices. They are provided for convenience by English readers).

15. Regulations for the Supervision and Administration of Medical Devices, State Council Order No. 276. China Food and Drug Administration. Available at http://eng.sfda.gov.cn/WS03/CL0767/61641.html; State Council Passes Draft Amendments to Device Regulations. China Food and Drug Administration. Available at http://eng.sfda.gov.cn/WS03/CL0757/96865.html.

16. Initial Registration for Overseas Medical Devices that are Approved for Marketing at Abroad. China Food and Drug Administration. Available at http://eng.sfda.gov.cn/WS03/CL0770/98145.html.

might be affected by any change in its design, raw material, manufacturing procedure, and so forth. In practice, the CFDA's local counterparts have wide discretion in determining what changes constitute material changes.

An immaterial/minor change in a registered medical device is no longer subject to the approval by the original registration authority; filing such change with the authority is sufficient.

Following the issuance of the new regulation, updates to the classification catalog of medical device types are likely. This may reposition devices into lower-risk categories and less-burdensome regulatory requirements.

Another significant change under the new regulation is that the term/validity of a medical device registration is 5 years instead of 4. Also, the registration can be extended through renewal, instead of re-registration. This should lower some of the regulatory burdens for manufacturers.

Clinical Trials

The 2004 SFDA rule allowed for the submission of clinical data from outside China for a number of devices.[17] Under the new regulations, clinical trial reports shall be issued by qualified clinical trial institutions located in China. Further, a clinical trial of low-risk medical devices no longer requires the approval of CFDA or its counterparts, but a filing is required. Certain Category III medical devices that involve very high risks must still obtain the necessary approvals.

The new regulations continue the requirement for clinical trials for both imported and locally manufactured Class II and III devices. A clinical trial may not be necessary if similar devices have been on the market for a long time and there is a proven record of safety, when there are no reports of serious adverse events. Additionally, clinical trials may not be needed if the safety and effectiveness of the product can be proven without a clinical trial. The CFDA will develop a catalog listing devices that will not require a clinical trial and the list will be made public.

The new regulations may contribute to greater regulatory efficiency, allowing a device company to first get its medical device registered, and then proceed to obtain the manufacturing license.

Medical devices manufactured in foreign countries must have their Chinese labels attached before they are imported into China. The new regulations contain a section on monitoring adverse incidents, reevaluating registered medical devices, and recalling defective medical devices. Every medical device enterprise must now establish an adverse incidents monitoring system to monitor and report adverse incidents. While some of these requirements are not new,[18] they underscore the CFDA's commitment to a regulatory system designed to improve the safety and quality of medical devices and health care in China.

17. Provisions for Clinical Trials for Medical Devices, SFDA Order No. 5. China Food and Drug Administration. Available at http://eng.sfda.gov.cn/WS03/CL0768/61644.html.

INCREASED LEGAL PENALTIES AND LIABILITIES

The old regulations imposed fines for certain serious violations based on two to five times of the illegal proceeds. Under the new regulations, however, it would be 5-10 times (or even 10-20 times) of the value of the involved medical devices.

The new regulations impose punishment for deceptive advertising of medical devices, and punishment for forgery, alteration, purchase, sale, and lease of any medical device certificates or permits.

CFDA is developing a number of important rules, normative documents, and guidelines that will implement many of these changes. These should clarify what manufacturers and distributors need to do to comply with the new regulations.

In addition to these new regulations, on May 15, 2014, the CFDA released its *Measures on the Supervision and Administration of the Quality of Medical Devices in Use* for public comment.[19] Under the measures, medical device operators will be required to establish a quality management system especially for Class III devices. Features of this proposed system cover the purchase of medical devices, an incoming stock inspection and recording system, an inbound and outbound management system, a daily maintenance and recording system, a quality traceability recording system, a management system for disposable medical devices, and a management system for contracts and technical documents for products.

PRACTICAL TIPS

Because China is a crucial part of the global market for innovative devices and drugs, we suggest some practical market access tips that anticipate China's complex regulatory environment.

1. Select the drugs or devices that address high-priority Chinese public health needs or niche markets.
2. Consider products already approved by the FDA and with a robust base of scientific data.
3. Include in your U.S. clinical trials and global clinical trial planning a statistically significant number of Asians to validate the efficacy of the drug/device on Asian/Chinese patients.
4. Perform clinical trials in China in those approved medical institutions with clinical experts that will use your product. Integrate China clinical trials with global regulatory strategies.
5. Start contacts with CFDA regulators at the national level very early in product planning.

---

18. CFDA releases 2013 Annual Report for National Medical Device Adverse Event Monitoring. China Food and Drug Administration. Available at http://eng.sfda.gov.cn/WS03/CL0757/99620.html.
19. *See* China Food and Drug Administration on "medical devices quality supervision and management approach (draft)" for public comment notice. China Food and Drug Administration. Available at http://www.sfda.gov.cn/WS01/CL0779/99834.html. *See also* CFDA Promulgates 120 Industry Standards for Medical Devices. China Food and Drug Administration. Available at http://eng.sfda.gov.cn/WS03/CL0757/102298.html (CFDA has also introduced 120 industry standards for medical devices).

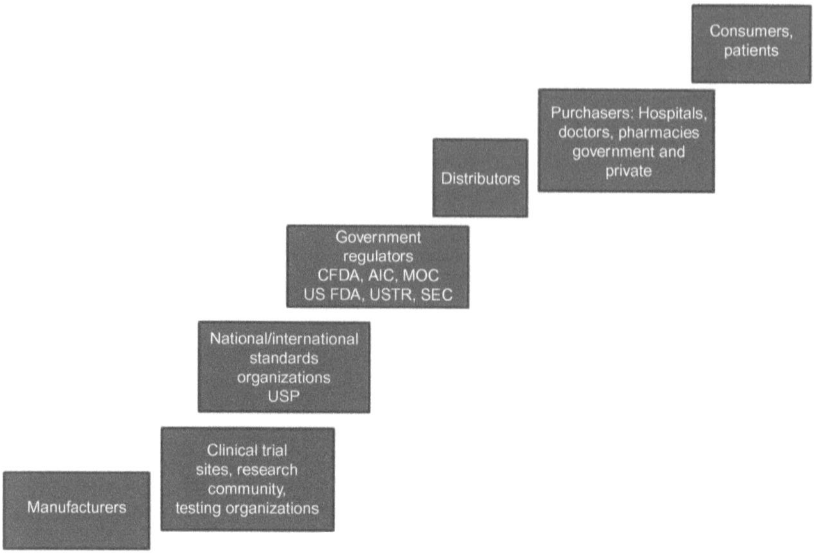

FIGURE 8.4 Key players in China US drug/device regulation.

6. Plan to work with CFDA review staff who may not be familiar with the underlying scientific and clinical bases for the drug or device. Arrange regular meetings with review staff to minimize the likelihood of expert panel review.
7. Establish a best-case and worst-case product approval schedule for importing, registering, and getting through CFDA. Be sure to include the need to re-register.
8. Build working relations with Chinese manufacturers and distributors.
9. Be careful with intellectual property and cybersecurity risks.
10. Make sure contacts with Chinese hospitals and doctors, directly and indirectly comply with Chinese and U.S. laws.
11. Plan for sales and procurement, especially in government hospitals that are obscure, difficult to access, and that compete with local Chinese companies.

China is moving methodically toward participation in global regulatory harmonization, which will be increasingly important in the coming years. Product review and approvals continue to be unpredictable and protracted. Companies can improve their market access when they can adapt to China's distinct regulatory requirements and national commitment to Chinese companies and a harmonious Chinese economy. Knowing these challenges can help multinational drug and device companies fully participate in this very promising market (Figure 8.4).

ACKNOWLEDGMENTS

The author thanks Judy Zakreski and Hoi Bun Suen, from China Trade Strategies, for their input on this chapter.

Part II

Designing Optimal Regulatory Structures to Address Diversifying and Globalizing Sources of Food and Additives

Chapter 9

The Balance Between Regulation and Private-Sector Initiative in Securing Healthy and Sustainable Food

Eleanore Alexander, Derek Yach, Adriana Selwyn
The Vitality Institute, New York, New York, USA

Chapter Outline

INTRODUCTION

The current food system is not sustainable. Global diets have shifted and are now high in calories and meat, producing excessive and unsustainable environmental externalities. An additional 60% more food will be required to feed the expected population of 9 billion in 2050. Agriculture currently accounts for 20%-35% of greenhouse gas (GHG) emissions.[1,2] Heart disease is the leading

1. Food and Agriculture Organization, 2014. Sustainable agriculture. Retrieved from http://www.fao.org/post-2015-mdg/14-themes/sustainable-agriculture/en/.
2. West P. C., Gerber, J. S., Engstrom, P. M., Mueller, N. D., Brauman, K. A., Carlson, K. M., Cassidy, E. S., Johnston, M., MacDonald, G. K., Ray, D. K., Siebert, S., 2014. Leverage points for improving global food security and the environment. Science 345, 325-328.

Food and Drug Regulation in an Era of Globalized Markets. http://dx.doi.org/10.1016/B978-0-12-802311-2.00009-6

cause of death globally, accounting for 30% of deaths in 2008.[3] Resulting from poor diet, physical inactivity, and tobacco use, cardiovascular disease is largely preventable. Obesity and related diseases have increased to all-time high levels; overweight and obesity affect 35% of the global population.[4] To address the current challenge of malnutrition (including over- and undernutrition), diets need to focus on items that are better for both population and environmental health.

The challenge of feeding a growing population while improving nutrition quality and reducing environmental impacts is complex and requires active participation by many: governments, industry, nongovernmental organizations (NGOs), and consumers. Partnerships and policies across sectors are needed to support agriculture, nutrition, and health. Placing health at the center of the development agenda would guarantee that national progress will protect and promote health and that population needs will be met through sustainable health systems.

HISTORICAL CONTEXT

Historically, public health has focused on government-led policies: taxes, regulations, educational campaigns, setting norms and standards, research support, and environmental change. These policies have served us very well for infectious diseases and many aspects of road safety, food safety, and tobacco control.

Tobacco control has been successful because the Framework Convention on Tobacco Control (FCTC) has held governments accountable via an international treaty by the World Health Organization (WHO).[5] This occurred after 50 years of policy development to reduce tobacco use and consequent diseases. Adopted by the WHO in 2003, the FCTC currently has 168 signatories, most recently El Salvador, and 179 parties.[6] Key strategies to build the FCTC included building a team to drive change, highlighting the role of the tobacco industry in preventing progress, supporting development of an NGO movement heeded by the media, gathering support for the US and Bretton Woods systems, building agreement with country-specific strategies, improving research through surveillance systems, connecting with the private sector including the sports industry, and supporting the best negotiators to lead the FCTC's progress.[7] Progress has been strong since the adoption of the treaty: 68% of the parties have implemented the

3. World Health Organization, 2014. Cardiovascular diseases. Retrieved from http://www.who.int/mediacentre/factsheets/fs317/en/.
4. World Health Organization, 2014. Obesity: situation and trends. Retrieved from http://www.who.int/gho/ncd/risk_factors/obesity_text/en/.
5. Yach, D., 2014. The origins, development, effects, and future of the WHO Framework Convention on Tobacco Control: a personal perspective. Lancet 383, 1771-1779.
6. World Health Organization, 2014. Parties to the Framework Convention on Tobacco Control. Retrieved from http://www.who.int/fctc/signatories_parties/en/.
7. Yach, D., 2014. The origins, development, effects and future of the WHO Framework Convention on Tobacco control: a personal perspective. Lancet 383, 1771-1779.

provisions, and more than 75% of parties strengthened existing legislation or adopted new policies following ratification.[8]

CHALLENGE: SUPPORTING HEALTHY DIETS AND ENVIRONMENTAL SUSTAINABILITY

Healthy diets and environmental sustainability are complex and interlinked issues. Food and nutrition policies have thus far had limited success in addressing these challenges. There are numerous reasons for this: health interests alone rarely control or even lead the approach to the issues, the goals of nutrition policy are rarely framed holistically, many countries pass laws without being able to enforce them, industry interests and lobby groups thwart development of coherent policies, governments severely underinvest in research to build broad-based solutions, and the potentially positive role of the private sector has been undermined by historic distrust. Finally, our models of behavior change have ignored advances in behavioral economic theory.

NARROW SCOPE OF NUTRITION POLICY

The goals of nutrition policy are rarely holistic but tend to address only small pieces of the puzzle such as food labeling, marketing, and nutrient levels. Transparent food labeling to improve consumer purchases is an increasingly widespread approach, such as in the United Kingdom where traffic light labels are used to communicate levels of salt, fat, saturated fat, and sugar in products. However, food labeling is complex and lacks consistency, in part due to discrepancies between governments viewing labeling as regulatory, and industry feeling that labeling should be voluntary and self-regulated.

Taxation has been considered as another approach at local and national levels to improve healthy consumer behaviors. In Mexico, taxes on soft drinks at 1 peso per liter and high-calorie foods at 8% of cost have been introduced. Tackling tobacco has been successful with policy action involving higher taxes. Tobacco policies, however, are not applicable to food issues because of the complex nature of food and eating: everyone needs to eat, and there are no agreed-on "good" foods and "bad" foods. For example, processed foods have been villainized; however, frozen and canned vegetables, which fall into this category, are often as nutritious as or more nutritious than fresh produce and often less expensive.[9]

8. WHO Tobacco Free Initiative, 2012. WHO report on the global tobacco epidemic, 2013: Enforcing bans on tobacco advertising, promotion and sponsorship. Retrieved from http://apps.who.int/tobacco/global_report/2013/en/.

9. American Heart Association, 2014. Fresh, frozen or canned fruits and vegetables: all can be healthy choices! Retrieved from http://www.heart.org/HEARTORG/GettingHealthy/NutritionCenter/HealthyCooking/Fresh-Frozen-or-Canned-Fruits-and-Vegetables-All-Can-Be-Healthy-Choices_UCM_459350_Article.jsp.

Tackling poor diet with taxes may encourage food substitution; for example, taxing sugar-sweetened beverages may reduce demand for soda but increase demand for products such as fruit juice that are also high in calories.[10] If we start to tax unhealthy foods, what other items should be included or excluded? Nuts, agreed on as a healthy food, can be very high in calories if overconsumed: a half-cup of almonds provides more than 400 calories. Nutrition policies, therefore, require comprehensive approaches and cannot rely on single measures.

LACK OF INVESTMENT IN BROAD-BASED SOLUTIONS

To improve diets while supporting the environment, malnutrition and nutrient quality must be addressed to ensure appropriate calorie and micronutrient intakes. Barry Popkin described the nutrition transition in detail: countries have shifted from dietary patterns of collecting food and eating plant-focused diets to diets high in fat and sugar coupled with reduced activity, leading to proliferation of noncommunicable diseases (NCDs).[11] In Mexico, 50% of calories come from packaged foods and beverages, and in China nearly 30% of calories come from packaged foods and beverages.[12] Worldwide, the top 10 packaged-food companies account for 15.2% of packaged-food sales, and each company contributes less than 3.3% of sales.[13] In the United States, Kraft holds the largest share, with 6.7% of packaged-food sales, and in South Africa Tiger Brands leads, with 17.2% of sales.

Addressing the needs of the 1 billion hungry, 1 billion micronutrient deficient, and 1 billion overweight requires increasing sustainable production of foods high in nutrient quality.[14] Major challenges outlined by the UK Foresight team on the Future of Food and Farming report include balancing future demand and supply to ensure affordability of foods, ensuring stability in food supplies, ensuring food security for all, addressing environmental impacts of food production, and maintaining biodiversity.[15] In addition, the response to global malnutrition must link to agriculture and support environmental sustainability.

On a global level, the response to other NCDs has been weak and disjointed.[16] Funding for NCDs is limited, comprising only 8% of the WHO budget

10. STATS, 2014. The problem with modeling soda taxes: price elasticity. Retrieved from http://www.stats.org/sodapage7.html.

11. Popkin, B. M., 2006. Global nutrition dynamics: the world is shifting rapidly toward a diet linked with noncommunicable diseases. American Journal of Clinical Nutrition 84, 289-298.

12. Popkin, B. M., 2014. Nutrition, Agriculture and the global food system in low and middle income countries. Food Policy 47, 91-96.

13. Alexander, E., Yach, D., Mensah, G. A., 2011. Major multinational food and beverage companies and informal sector contributions to global food consumption: implications for nutrition policy. Global Health 7(26). Retrieved from http://www.globalizationandhealth.com/content/7/1/26.

14. Worldwatch Institute, 2011. The Forgotten 1 billion. Retrieved from http://www.worldwatch.org/forgotten-1-billion.

15. Government Office for Science, 2011. The future of food and farming: challenges and choices for global sustainability. Retrieved from https://www.gov.uk/government/uploads/system/uploads/attachment_data/file/288088/11-547-future-of-food-and-farming-summary.pdf.

16. Gostin, L. O., 2014. Healthy living needs global governance. Nature 511, 147-149.

compared with 39% for infectious diseases and polio eradication. Support for NCDs is limited, as seen in the late time frame for a high-level meeting and in the lack of activism and action during and after the meeting. NCDs need to be reframed as a collective problem warranting a global response.[17] This includes ensuring sustainable funding to build capacity in low-income countries, regulating the food industry to ensure effective change in countries, investing in the built environment to support public transportation and physical activity, and prioritizing prevention to coordinate ministers from multiple sectors of government or international organizations.

RELATIONSHIP WITH THE PRIVATE SECTOR

It is simple to tax and ban tobacco, but successfully improving diets requires a more nuanced approach. This approach must promote healthy foods, eliminate marketing of unhealthy products to children, inform and support portion control, and address the dangers and stigma of obesity. Because of the complex and multifaceted approach needed to improve diets, we must consider alternatives, including public-private partnerships (PPPs). Leaving the market to its own devices will not meet the challenge.

The political declaration by member states at the UN high-level meeting endorsed a whole of society, whole of government approach and multistakeholder action to prevent and control NCDs. Other UN agencies are also more supportive of PPPs and the private sector, as seen by the Food and Agriculture Organization (FAO) that "considers the Private Sector to be a key ally in the fight against hunger."[18] The UN Economic and Social Council and FAO are actively working with the International Food and Beverage Alliance (IFBA), a group of leading global food and nonalcoholic beverage companies committed to helping consumers achieve balanced diets and healthy, active lifestyles. The IFBA has made progress on four key areas: product formulation and innovation, consumer information, responsible advertising and marketing to children, and promotion of healthy lifestyles. Since 2004, IFBA companies have been steadily reducing or removing ingredients of public health concern—salt, sugar, saturated fat, and trans fat—and increasing beneficial ingredients like fiber, wholegrains, fruits and vegetables, and low-fat dairy. Member compliance with the IFBA Global Policy on Advertising and Marketing to Children has been above 86% for television and 100% for print and Internet advertising.[19]

17. Gostin, L. O., 2014. Healthy living needs global governance. Nature 511, 147-149.
18. Food and Agriculture Organization of the United Nations, 2014. FAO Considers the Private Sector to be a Key Ally in the Fight against Hunger. Retrieved from http://www.fao.org/partnerships/private-sector/en/.
19. International Food and Beverage Alliance, 2014. Highlights of Our Progress 2008-2013. Retrieved from https://ifballiance.org/documents/2014/09/highlights-of-our-progress-2008-2013.pdf.

Though PPPs can have significant impact, distrust of industry among the public sector prevents successful partnerships with the private sector. In fact, the distrust exists even at the executive level of the WHO:

> *Efforts to prevent NCDs go against the business interests of powerful economic operators. In my view, this is one of the biggest challenges facing health promotion... it is not just Big Tobacco anymore. Public health must also contend with Big Food, Big Soda, and Big Alcohol. All of these industries fear regulation, and protect themselves by using the same tactics... This is formidable opposition. Market power readily translates into political power. Few governments prioritize health over big business. As we learned from experience with the tobacco industry, a powerful corporation can sell the public just about anything.*[20]
>
> —Margaret Chan, WHO Director-General

Although distrust exists in some in the public sector and NGO community, these partnerships have proven successful in supporting population and business health. Open dialog is key to sharing knowledge between the public and private sectors, who often view each other with skepticism or even aversion. While the public sector is interested in taking action to promote health, the private sector has an opportunity, with corporate input into accelerating initiatives, to achieve faster success with greater impact.[21]

LIMITED CAPACITY TO ENFORCE LAWS

Public health law is defined as "the legal powers and duties of organized society to assure the conditions of the people to be healthy."[22] However, more than legislation is needed for public health law to "exercise its functions."[23] Legislation is only as strong as its enforcement, and monitoring and evaluation of regulation is often weak due to lack of government capacity. Few governments worldwide have the expertise or funds to implement a comprehensive monitoring and evaluation system for responsible marketing to children or product reformulation. In addition, WHO and government do not necessarily accept the independent monitoring programs funded by industry and would prefer such programs to be led by nonprofits and academics who might have significant conflicts of interest themselves. Large companies can be forced into changing practices by public pressure to maintain their reputation internationally, but the informal food sector, very large in many countries, falls outside of regulation-enforcement capabilities. If regulations are enacted to reduce sodium intake, for example, the informal sector has no obligation—and likely no inclination—to reduce the sodium content of their products

20. Chan, M., 2013. WHO Director-General addresses health promotion conference. World Health Organization. Retrieved from http://www.who.int/dg/speeches/2013/health_promotion_20130610/en/.
21. Yach, D., 2014. Food industry: friend or foe? Obesity Review 15, 2-5.
22. Gostin, L. O., 1996. Public health law: a review. Current Issues in Public Health 2, 205-214.
23. L'hirondel, A., Yach, D., 1998. Develop and strengthen public health law. World Health Statistics Quarterly 51, 79.

if it reduces the taste or increases the expense of production.[13] Few studies have investigated the informal sector, but it has been found that in Bangkok, Thailand, 40% of energy intake, and up to 88% of energy intake by children, is from the informal sector.[24] When enforcement for all food products does not exist, regulations become very challenging, as large producers must change their products while informal producers likely will not, and consumers who do not like the taste of the major-brand products that adhere to regulation have an easy opportunity to purchase less-nutritious products from the informal sector.

THE WAY FORWARD

Opportunities exist to improve the nutritional quality and the environmental impact of the current food system. The Global Landscapes Initiative team identified global leverage points offering opportunities to improve food security and environmental sustainability; for example, by increasing total production to 358 megatons per year and closing the yield gap to 50% of attainable yields.[25] The Global Landscapes Initiative team calculated that about 70% more calories would become available if crops currently used for animal feed and nonfood uses, including biofuels, were targeted for direct consumption. These additional calories could support an additional 4 billion people. Current GHG emissions are mostly caused by tropical deforestation, livestock and rice production, and fertilizers. These emissions largely occur in a few countries and from specific crops. Targeting these countries and crops can reduce GHG emission significantly.

FOSTER NEW MODELS OF COLLABORATION

The World Bank defines PPPs as "arrangements, typically medium to long term, between the public and private sectors whereby some of the services that fall under the responsibilities of the public sector are provided by the private sector, with clear arrangement on shared objectives for delivery of public infrastructure and/or public services."[26] PPPs can be successful if the strengths of each partner are leveraged to create greater benefit than either partner could achieve alone. However, clear governance must be established, as in any other partnership, to ensure that all partners contribute toward shared success. Clear governance and transparency help assure the public that the PPP is legitimately serving the stated goals.

---

24. Rural Infrastructure and Agro-Industries Division, Nutrition and Consumer Protection Division of the Agriculture and Consumer Protection Department of FAO, 2007. Promises and challenges of the informal food sector in developing countries. Retrieved from http://www.fao.org/docrep/010/a1124e/a1124e00.htm.

25. West, P. C., Gerber, J. S., Mueller, N. D., Brauman, K. A., Carlson, K. M., Cassidy, E. S., Engstrom, P. M., Johnston, M., MacDonald, G. K., Ray, D. K., Siebert, S., 2014. Leverage points for improving food security and the environment. Science 345, 325-328.

26. World Bank, 2014. What are public private partnerships? Retrieved from http://ppp.worldbank.org/public-private-partnership/overview/what-are-public-private-partnerships.

Partnerships, including the Healthy Weight Commitment Foundation (HWCF),[27] the Partnership for a Healthier America,[28] the IFBA, and those developed through the Clinton Global Initiative (CGI),[29] show that collaboration can help push companies to produce healthier products.

- The HWCF reduced 6.2 trillion calories in the marketplace from 2007 to 2012, surpassing by more than 400% the pledge to remove 1.5 trillion by 2015.[30] This pledge was independently evaluated through funding by the Robert Wood Johnson Foundation.
- The Partnership for a Healthier America, an organization collaborating with the private sector to address childhood obesity, has more than 50 partners collaborating on obesity prevention.[31] Birds Eye Foods committed to a $6 million marketing campaign to promote vegetables by partnering with Nickelodeon's series *iCarly*.[32]
- A partnership led by the Alliance for a Healthier Generation brought together the American Heart Association, the Clinton Foundation, and the American Beverage Association to change beverages available in schools to lower- and no-calorie options. Following this pledge there was an 88% decline in calories of beverages shipped to schools.[33]
- IFBA members have committed to
  - Reformulate products and develop new products that support the goals of improving diets
  - Provide clear and fact-based nutrition information to all consumers
  - Extend responsible advertising initiatives and marketing to children globally
  - Raise awareness of balanced diets and increased levels of physical activity

---

27. Healthy Weight Commitment Foundation, 2014. Working Together to Change the Outlook of a Generation. Retrieved from http://www.healthyweightcommit.org/images/uploads/HWCF_5_Report_2014/HWCF_5YearAnniversary_FW.pdf.
28. Partnership for a Healthier America, 2013. In it for Good: 2013 annual progress report. Retrieved from https://drive.google.com/file/d/0B-6Ewy5pv-vZQzgzLVd6Yl91dWM/edit.
29. American Beverage Association, 2014. Alliance for a Healthier Generation and America's Beverage Companies Announce Landmark CGI Commitment to Reduce Beverage Calories Consumed Across the Nation. Retrieved from http://www.ameribev.org/news-media/news-releases-statements/more/334/.
30. Healthy Weight Commitment Foundation, 2014. Major Food, Beverage Companies Remove 6.4 Trillion Calories from U.S. Marketplace. Retrieved from http://www.healthyweightcommit.org/news/major_food_beverage_companies_remove_6.4_trillion_calories_from_u.s._market/.
31. Partnership for a Healthier America, 2014. Our Partners. Retrieved from http://ahealthieramerica.org/our-partners/.
32. PR Newswire, 2012. Birds Eye, America's Leading Vegetable Company, Joins Partnership for a Healthier America in the Fight Against Childhood Obesity. Retrieved from http://www.prnewswire.com/news-releases/birds-eye-americas-leading-vegetable-company-joins-partnership-for-a-healthier-america-in-the-fight-against-childhood-obesity-150420245.html.
33. American Beverage Association, 2012. Journal Article Validates Beverage Industry's Success in Removing Full-Calorie Soft Drinks from Schools Nationwide. Retrieved from http://www.ameribev.org/nutrition-science/school-beverage-guidelines/news-releases/more/284/.

- Actively support PPPs that adopt the WHO's 2004 *Global Strategy on Diet, Physical Activity and Health*
- The CGI promotes partnership for action.[34] With a mission to turn ideas into action, the CGI holds annual meetings to bring leaders together to address global challenges through action-focused commitments. More than 3100 commitments have been made by the CGI community, impacting more than 430 million people in 180 countries. At the CGI annual meeting, the Reinvestment Fund announced its intentions to invest $100 million in health-food businesses over the next 3 years.[35]

Whereas critics assume that food companies can only succeed by selling unhealthy foods, a recent report by the Hudson Institute shows that healthier products support corporate growth.[36] Lower-calorie items now contribute 52.5% of sales by HWCF companies, an increase from 50.5% in 2012. In addition, lower-calorie items drove 99% of sales growth over a 5-year period. Motivating factors for industry efforts to improve diet have included progressive chief executive officers (CEOs) taking action, investors rewarding companies considering health as part of a longer-term sustainability strategy, and increased consumer demand.[37] In addition, the Bank of America/Merrill Lynch Globesity report points out that efforts to address obesity are an investment "megatrend," and four categories of companies meeting the obesity challenge will be well positioned for long-term sustainability: pharmaceuticals and health care; food; commercial weight loss, diet management, and nutrition; and sports apparel and equipment.[38]

In addition to PPPs focusing on nutrition, partnerships in related areas offer insight into best practices. GAVI, a global vaccine alliance, and the World Economic Forum's New Vision for Agriculture are examples of successful and impactful partnerships.

- GAVI is an international PPP focused on providing equal access to new and underused vaccines for children.[39] GAVI partners include the Bill and Melinda Gates Foundation, WHO, the United Nations Children's Fund, and the World Bank. Each partner's unique strengths contribute to the partnership for funding, implementation, vaccine expertise, and

---

34. Clinton Foundation, 2014. CGI Model. Retrieved from https://www.clintonfoundation.org/clinton-global-initiative/about-us/cgi-model.
35. The Reinvestment Fund, 2014. TRF to bring $100 million to healthy food businesses in underserved communities. Retrieved from http://www.trfund.com/trf-to-bring-100-million-to-healthy-food-businesses-in-underserved-communities/.
36. Hudson Institute, 2014. Lower-calorie foods and beverages fuel growth at Healthy Weight Commitment Foundation companies. Retrieved from http://s3.amazonaws.com/media.hudson.org/files/publications/HWCF_Hudson_Report__FINAL.pdf.
37. Yach, D, 2014. Can we leave industry to lead efforts to improve population health? Yes. The BMJ 346.
38. Bank of America Merrill Lynch, 2012. Efforts to Tackle Global Obesity Shaping a New Investment Megatrend, Says New BofA Merrill Lynch Report. Retrieved from http://newsroom.bankofamerica.com/press-release/economic-and-industry-outlooks/efforts-tackle-global-obesity-shaping-new-investment-me.
39. Gavi, 2014. About Gavi, the Vaccine Alliance. Retrieved from http://www.gavi.org/about/.

development assistance. GAVI has specific indicators to measure success. Recently, GAVI approved a framework to enable immunization of 300 million children between 2016 and 2020.[40]

- The New Vision for Agriculture started with development of a policy document that emphasized the need for better integration between agricultural productivity, environmental sustainability, and improved nutrition for all. The New Vision work has now evolved to focus on a few specific countries, particularly in Africa, where governments, companies that extend across the agricultural supply chain from seed to food, key NGOs, academia, and public and private investors are working together. The goal is to create conditions that attract investment and stimulate actions that will benefit local smallholder farmers and business and meet governments' needs for export earnings and multinationals' long-term needs for reliable and sustainable sources of affordable commodities.

EMBRACE THE SHIFT TOWARD LIBERTARIAN PATERNALISM

US consumers—as evidenced by millennials everywhere—are wary of infringement on their rights, even if it is done to promote the public's health. In New York City, former Mayor Michael Bloomberg was able to remove trans fats from the restaurant food supply in 2007-2008; however, additional modifications to the food supply to promote public health (such as the limit on sugar-sweetened beverage size) have been tough to achieve.[41] New York City has labeled calories on chain restaurant items since 2008, and in 2013 the Affordable Care Act required all US chain restaurants to label menus.[42] Although evidence for menu labeling having an effect on consumer choice is mixed, the required labeling pushes restaurants to reformulate products to reduce calories.[43]

Because of this fear of paternalism and demand for choice, there has been a shift toward libertarian paternalism. Defined by Cass Sunstein and Richard

40. Gavi, 2014. GAVI Alliance Board approves new strategic framework to reach an additional 300 million children with vaccines. Retrieved from http://www.gavi.org/Library/News/Press-releases/2014/GAVI-Alliance-Board-approves-new-strategic-framework-to-reach-an-additional-300-million-children-with-vaccines/.

41. New York City Department of Health, 2008. Board of Health votes to phase out artificial trans fat from New York City's restaurants. Retrieved from http://www.nyc.gov/html/doh/html/pr2006/pr114-06.shtml.

42. Federal Register, 2014. Food Labeling: Nutrition Labeling of Standard Menu Items in Restaurants and Similar Retail Food Establishments. Retrieved from https://www.federalregister.gov/regulations/0910-AG57/food-labeling-nutrition-labeling-of-standard-menu-items-in-restaurants-and-similar-retail-food-estab.

43. Robert Wood Johnson Foundation, 2013. Impact of Menu Labeling on Consumer Behavior: A 2008-2012 Update. Retrieved from http://www.rwjf.org/content/dam/farm/reports/reports/2013/rwjf406357.

Thaler, libertarian paternalism is when "private and public institutions might nudge people in directions that will make their lives go better, without eliminating freedom of choice."[44] Examples of this are setting healthy defaults such as serving a side salad instead of french fries and using smaller plates to encourage eating less. The proposal by Mayor Bloomberg to limit the maximum size of fountain beverages to 16 oz. is another example, as it allowed a choice of size and did not limit the amount of beverages individuals could purchase. However, this proposal was heavily lobbied against by the beverage industry and was rejected by the New York State Court of Appeals.

ENHANCE THE ROLE OF GOVERNMENT

In Australia, a new public health governance model for reducing population salt intake has been proposed. The model is based on a responsive regulatory approach whereby failure by the food industry to meet salt-reduction targets results in progressively more stringent regulation. The three-phase strategy involves selective introduction of "legislative scaffolds" to accelerate salt reduction through product reformulation while seeking to maximize industry cooperation and innovation. In phase 1, food manufacturers have the opportunity to negotiate targets for reformulating foods. In phase 2, companies that failed to reformulate noncomplying products would be required to include mandatory warning labels indicating that the product is high in salt. This would be in parallel with mass media and social marketing campaigns encouraging consumers to lower salt intake; such a penalty would therefore be a significant driver for compliance. In addition, compliance data would be made publicly available. Finally, in phase 3, court-enforced penalties would be applied to manufacturers who continue to undermine salt-reduction efforts.[45]

This approach gives the food industry the opportunity to demonstrate leadership on a voluntary basis, yet also highlights the role government can play in accelerating progress. Governments can take a number of steps to encourage industry compliance with voluntary regulations by increasing market pressure for regulated companies and linking publicity campaigns to improved consumer understanding of these schemes. The UK's salt-reduction initiative included a government-led education campaign to improve consumer knowledge on the link between salt and health and to encourage reduction in salt intake. The UK Food Standards Agency played a key role in encouraging front-of-pack traffic light labeling on food products, which helped drive reformulation and development of more healthy products.

---

44. University of Chicago, 2007. Libertarian paternalism. Retrieved from http://uchicagolaw.typepad.com/faculty/2007/01/libertarian_pat.html.
45. Magnusson, R., Reeve, B., 2014. 'Steering' private regulation? A new strategy for reducing population salt intake in Australia. Sydney Law Review 36, 255-289.

Governments need to increase research efforts and educational campaigns to further progress in changing the food supply. New and existing policies require greater coherency. For example, integrating US government food assistance programs such as the Supplemental Nutrition Assistance Program (SNAP) and the Special Supplemental Nutrition Program for Women, Infants and Children (WIC) could boost the impact of these programs.

LEVER CORE CAPABILITIES OF COMPANIES

Companies have expertise that can be shared to help promote optimal nutrition globally. Areas in which food companies can address undernutrition through business practices include investing in agriculture, expanding the use of corporate distribution capabilities, providing greater support for food and beverage fortification, innovating and expanding foods for undernutrition, establishing innovative social business models, investing in development of nutrition science capacity, reformulating low-cost nutritious foods, and advocating nutrition-friendly trade policies.[46] To improve diet, physical activity, and health, the WHO guides industry to promote healthy diets and physical activity; limit fat, sugar, and salt in products; develop and provide healthy and affordable products; provide adequate and understandable nutrition information; market responsibly, especially to children; use clear and consistent labeling and evidence-based claims; share food-composition information with the government; and support development and implementation of physical activity programs.[47]

Companies have strong skills in behavioral sciences because of their focus on understanding and influencing consumers, which can be applied to public health. Specifically, marketing via consumer communication is an area of expertise of companies. In addition, the investment in R&D is very strong in companies, although industry-funded research is often not considered credible. Rather than focusing on the conflict of interest of industry when every organization has a vested interest, peer review should be used to guarantee independence.[48]

Alignment of agricultural practices with attainment of optimal nutrition for all is crucial. Improving agriculture for better nutrition requires farmers to improve production and must be supported by business and government.[49] Government can offer incentives for growing nutritious crops and businesses can support infrastructure for healthier products. Countries can look to regional successes that can be translated into improved population health and nutrition status through

---

46. Yach, D., Feldman, Z. A., Bradley, D. G., Khan, M., 2010. Can the food industry help tackle the growing burden of undernutrition? American Journal of Public Health 100, 974-980.
47. World Health Organization, 2004. Global strategy on diet, physical activity and health. Retrieved from http://www.who.int/dietphysicalactivity/strategy/eb11344/strategy_english_web.pdf.
48. Caplan, A.L., 2012. Is industry money the root of all conflicts of interest in biomedical research? Annals of Emergency Medicine 59, 87-88.
49. Yach, D., 2011. Agriculture for better nutrition. Huffington Post Blog. Retrieved from http://www.huffingtonpost.com/dr-derek-yach/agriculture-for-better-nu_b_969066.html.

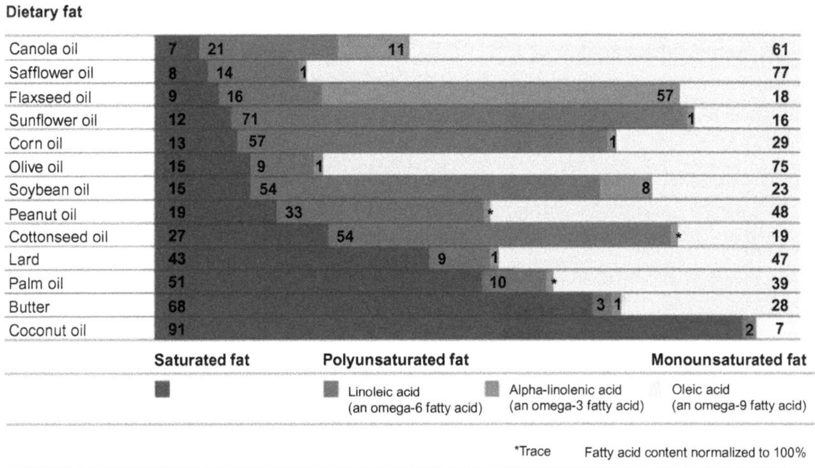

FIGURE 9.1 Comparison of dietary fats. *http://www.canolainfo.org/health/index.php?page=12.*

agriculture.[50] Further, corporate supply chain expertise can benefit public health, for example efficiently distributing vaccines and food in times of famine.

For example, PepsiCo has partnered with the World Food Programme (WFP) to share logistics and technical support with WFP colleagues, improving delivery of food and relief to WFP projects.[51] PepsiCo has also partnered with the Inter-American Development Bank to develop a local source of sunflower oil, supporting local farmers while improving the nutrition quality of snack products.[52] The Inter-American Development Bank will provide up to $5 million in credit guarantees to support sunflower production. PepsiCo has committed to spend $52 million on sunflower crops from 2011 to 2018 to produce 40,000 tons of sunflower oil. The sunflower oil will be used to replace palm oil, which has more than four times the saturated fat (see Figure 9.1).

BUILD ON 40 YEARS OF PARTNERSHIPS BETWEEN UNLIKELY PARTNERS

Only through **new and innovative partnerships** *can we make a difference.* [53]

—Gro Harlem Brundtland, former WHO Director-General

---

50. Fan, S., Pandya-Lorch, R. (Eds.), 2012. Reshaping agriculture for nutrition and health. International Food Policy Research Institute. Retrieved from http://www.ifpri.org/publication/reshaping-agriculture-nutrition-and-health.

51. World Food Programme, 2014. PepsiCo: Logistics to save lives. Retrieved from http://www.wfp.org/partners/private-sector/meet-our-partners/pepsico.

52. Strom, S., 2011. For Pepsi, a business decision with social benefit. New York Times. Retrieved from http://www.nytimes.com/2011/02/22/business/global/22pepsi.html?pagewanted=all.

53. Brundtland, G. H., 2002. Address by Dr. Gro Harlem Brundtland, Director-General, to the Fifty-Fifth World Health Assembly. World Health Organization, Geneva.

Gro Harlem Brundtland speaks to the fact that "business as usual" is shifting to innovative models that create financial success with social impact. For example, Cornerstone Capital, an institutional investment and financial advisory firm, is a leader in building partnerships to promote sustainability. The firm's mission is to bring the discipline of sustainable finance to the mainstream of the capital markets. Led by Erika Karp, Cornerstone wants to "accelerate the velocity of money... what is finance supposed to do? It's supposed to make capital available, and once you do that, theoretically there should be a multiplier effect in the economy. We don't have money moving fast enough."[54] The idea came in part from the World Economic Forum's Global Agenda Council on Financing and Capital, a group that seeks to address how capital and financing can be allocated in line with economic growth and minimal adverse effects.

RETURN MOBILITY TO CITIES THROUGH INNOVATIVE INFRASTRUCTURE INVESTMENTS

Investing in city infrastructure to support physical activity and active transport is critical to ensuring livable cities with healthy and successful populations.

The New York City High Line is a strong example of transforming cities to support physical activity and healthy living.[55] The High Line is a former elevated-train line that was transformed into a public park; it had 4.8 million visitors in 2013.[56]

Another example of urban infrastructure to support health is the Atlanta BeltLine, a transportation and economic development effort linking public parks, trails, and transit along a 22-mile railroad track.[57] The BeltLine includes 22 miles of pedestrian-friendly rail transit, 33 miles of multiuse trails, 1300 acres of parks, 5600 units of affordable housing, 1100 acres of remediated brownfields, and public art.

THE FUTURE

To address the current challenge of sustainably feeding a growing population, all sectors must take action to support a healthier people and planet. Smart laws that promote innovation in science and business aimed at achieving sustainable development are needed to support individuals and businesses. The private sector must develop products that support health with limited environmental impact. The biggest trade-off companies must address is between selling products,

54. Roston, E., 2013. Former UBS banker wants to 'accelerate the velocity of money'. Bloomberg. Retrieved from http://www.bloomberg.com/news/2013-08-07/former-ubs-banker-wants-to-accelerate-the-velocity-of-money-.html.

55. High Line, 2014. Retrieved from http://www.thehighline.org/.

56. Grynbaum, M. M., 2014. High line draws millions, but de Blasio isn't one. Retrieved from http://www.nytimes.com/2014/10/02/nyregion/high-line-draws-millions-but-de-blasio-isnt-one.html.

57. Atlanta BeltLine, 2014. The Atlanta Beltline: The 5 Ws and Then Some. Retrieved from http://beltline.org/about/the-atlanta-beltline-project/atlanta-beltline-overview/.

popular in member countries of the Organization for Economic Co-operation and Development, that have a negative impact on sustainability and forfeiting or delaying profits while new products and business models are developed. For example, if Chinese per capita meat consumption approached that in the United States, the world's grain supply would be threatened, heart disease would continue to increase in China, and widespread environmental damage would occur.[58]

Though the challenge to provide healthy environmentally sustainable diets for the growing population is immense, several leaders highlight the opportunity for solutions. General Electric's work on reverse innovation in India-focused, for example, on low-cost, more-portable diagnostics and suggests that a more sustainable course is possible.[59] McKinsey analysis shows that opportunities are available to expand supply and increase productivity to address resource challenge constraints expected in the next 20 years.[60] Accenture reports that corporate sustainability contributes directly to business value, whether through revenue generation, cost control, risk management, or long-term value creation. Sustainability should therefore be a strategic issue for any organization.[61] Finally, companies that have been defined as "high" in creating shared value perform better on several financial measures than others with "low" categorization.[62] Examples include Nestlé, Unilever, and Novo Nordisk's triple bottom line.

Companies addressing sustainability show deeper board and senior management involvement in driving sustainability, better-organized approaches to continuously engaging external stakeholders, and better metrics of nonfinancial performance.[63]

Porter and Kramer emphasize the importance of redefining the corporation as creating shared value rather than profit alone.[64] Shared value "recognizes that societal needs, not just conventional economic needs, define markets." By redefining markets and products, improving productivity in the value chain, and

58. Yach, D., 2012. Business and Society Conference: "Trading-off: Impactful Business Strategy in Uncertain Times". Tuck School of Business.

59. Schmind, J., 2009. GE Healthcare advances global reach with new heart monitor. Journal Sentinel. Retrieved from http://www.jsonline.com/business/41240677.html.

60. Dobbs, R., Oppenheim, J., Thompson, F., Brinkman, M., Zornes, M., 2011. Resource revolution: meeting the world's energy, materials, food, and water needs. McKinsey & Company. Retrieved from http://www.mckinsey.com/insights/energy_resources_materials/resource:revolution.

61. Accenture, 2011. Sustainability performance management: how CFOs can unlock value. Retrieved from http://www.accenture.com/SiteCollectionDocuments/PDF/Accenture-Sustainability-Performance-Management.pdf.

62. Yach, D., 2012. Business and Society Conference: "Trading-off: Impactful Business Strategy in Uncertain Times". Tuck School of Business.

63. Eccles, R. G., Ionnaou, I., Serafeim, G., 2011. The impact of a corporate culture of sustainability on corporate behavior and performance. Harvard Business School. Retrieved from http://hbswk.hbs.edu/item/6865.html.

64. Porter, M. R., Kramer, M. R., 2011. Creating shared value. Harvard Business Review. Retrieved from http://hbr.org/2011/01/the-big-idea-creating-shared-value/ar/1.

allowing local cluster development, companies can drive economic value by creating shared value. Opportunities for shared value are increasing and can affect all company decisions.

CONCLUSION

Addressing the unsustainable food system is critical to reducing the health and environmental risks that are projected to increase unless significant changes are made. New models of collaboration are needed, including PPPs, to promote accelerated progress improving nutrition while supporting the environment. Successful PPPs offer models for future collaborations as well as evidence that the private sector is a necessary actor to include in the effort to achieve sustainable diets. Government can encourage progress with smart regulation, and companies can use their core capabilities to support health and sustainability for the population.

Chapter 10

Divide and Confound: The Relationship Between Transparency, Public Health, and Regulatory Authority in the National Food System

Lisa Heinzerling
Georgetown University Law Center, Washington, District of Columbia, USA

The renowned agrarian and essayist Wendell Berry often quotes Sir Albert Howard, a British agronomist and one of the founders of modern organic agriculture, as urging us to reconceive "the whole problem of health in soil, plant, animal and man as one great subject."[1] Berry believes that a defining problem of our food system is our failure to embrace Howard's holistic perspective:

> *[I]f we understand the farm as an organism, we see that it is impossible to sacrifice the health of the soil to improve the health of plants, or to sacrifice the health of plants to improve the health of animals, or to sacrifice the health of animals to improve the health of people. In a biological pattern—as in the pattern of a community—the exploitive means and motives of industrial economics are immediately destructive and ultimately suicidal.... What is good for the water is good for the ground, what is good for the ground is good for the plants, what is good for the plants is good for animals, what is good for animals is good for people, what is good for people is good for the air, what is good for the air is good for the water. And vice versa.*[2]

Harmony, not splintering, is Berry's watchword. If one looks across the U.S. food system, however, one finds much splintering and little harmony. Indeed, one of the great problems of the food system in the United States is its failure to understand and embody the unity of the health of the environment in which food is grown and the wholesomeness of the food itself.

1. Berry, W., 1990. The Pleasures of Eating, in What Are People For? North Point Press.
2. Berry, W., 1981. Solving for Pattern, in The Gift of Good Land: Further Essays Cultural & Agricultural. North Point Press.

Food and Drug Regulation in an Era of Globalized Markets. http://dx.doi.org/10.1016/B978-0-12-802311-2.00010-2

In law, no feature of our food system better symbolizes this failure than the scattering of regulatory authority over the food supply across multiple federal agencies with multifarious missions. The division of regulatory responsibility for food across multiple federal agencies in the United States is, among aficionados of food law, a continual source of consternation and befuddlement. Some 15 different federal agencies, from 6 different cabinet-level institutions, share authority over the safety and transparency of our food supply, together administering 35 different federal laws.[3] Nine different congressional committees play a role in directing and overseeing the food safety work of these agencies.[4] Fifty-one separate interagency agreements aim to tame the jurisdictional chaos.[5]

As arresting as these statistics are, they cannot match concrete examples in conveying the extreme strangeness of the allocation of authority over food safety and transparency in the United States. Specific examples of regulatory fragmentation abound, but several, justly, have become famous, so well do they capture the puzzling allocations of regulatory authority. Consider the division of authority over eggs alone: the Food and Drug Administration (FDA) regulates the safety of eggs *in* the shell, the Food Safety and Inspection Service (FSIS) of the U.S. Department of Agriculture (USDA) regulates the safety of eggs *removed* from their shells, and the USDA's Agricultural Marketing Service (AMS) grades the eggs (in their shells) and monitors the cleanliness of the facilities in which the eggs are laid.[6] Or ponder these head scratchers: the FDA oversees the safety of meat and poultry sandwiches—unless they're open-face, in which case authority shifts to the USDA; the FDA oversees hot dogs wrapped in a bagel-style cover, but the USDA oversees hot dogs covered in cornmeal batter; authority over frozen cheese pizza goes to the FDA, but if the pizza has a meat topping, authority over it moves to the USDA; the FDA gets the regulatory nod for bottled water, but the Environmental Protection Agency (EPA) gets

3. U. S. General Accounting Office, 2013. Report to Congressional Committees, GAO-13-283, High-Risk Series: An Update; Congressional Research Service, 2014. The Federal Food Safety System: A Primer; Nestle, M., 2002. Food Politics: How the Food Industry Influences Nutrition and Health. University of California Press.

4. Congressional Research Service, 2014. The Federal Food Safety System: A Primer.

5. U. S. General Accounting Office, 2004. Testimony Before the Subcommittee on Civil Service and Agency Organization, Committee on Government Reform, House of Representatives, Federal Food Safety and Security System: Fundamental Restructuring Is Needed to Address Fragmentation and Overlap; U. S. General Accounting Office, 1990. Report to Congressional Requesters, Food Safety and Quality: Who Does What in the Federal Government, GAO/RCED-91-19A.

6. Hawthorne, F., 2005. Inside the FDA: The Business and Politics Behind the Drugs We Take and the Food We Eat. John Wiley & Sons, Inc.

it for tap water.[7] Even President Obama has joined in poking fun at this regulatory hodgepodge. In a segment on regulatory inefficiencies and excesses in his State of the Union address in 2011, the president observed, to predictable laughter and applause:

> *The Interior Department is in charge of salmon while they're in fresh water, but the Commerce Department handles them when they're in saltwater. I hear it gets even more complicated once they're smoked.*[8]

Amusing as it is, as a sendup of mindless bureaucratic distinctions, the president's message also exposes serious problems in our regulatory system for food.

Perhaps the most obvious problem with regulatory fragmentation is inefficiency, with its attendant waste of scarce resources. Understandably, the agencies in this tangled regulatory system do not always precisely organize their tasks to avoid duplication and overlap. For example, different agencies frequently inspect the very same facilities for the same kinds of problems; in fact, 6% of the FDA's inspections of food processing facilities in six states from 1987 to 1991 duplicated inspections performed by other agencies.[9] The system cannot afford this kind of overlap. Inspection of food processing facilities is one of the core legal responsibilities of agencies overseeing the food system, yet resources for inspections have in the last several decades suffered alarming declines even as legal mandates for inspections have grown. The FDA has never recovered the level of inspection resources (or the level of actual inspections) that it had before the Reagan administration's deregulatory push in the 1980s.[10] More recently, much has been made of the substantive regulatory improvements ushered in by the Food Safety Modernization Act (FSMA), but less has been said about the law's failure to provide anything close to adequate resources for

---

7. U. S. General Accounting Office, 2004. Testimony Before the Subcommittee on Civil Service and Agency Organization, Committee on Government Reform, House of Representatives, Federal Food Safety and Security System: Fundamental Restructuring Is Needed to Address Fragmentation and Overlap; U. S. General Accounting Office, 2001. Testimony Before the Subcommittee on Oversight of Government Management, Restructuring and the District of Columbia, Committee on Governmental Affairs, U.S. Senate, Food Safety and Security: Fundamental Changes Needed to Ensure Safe Food; Taylor, M. R., 1997. Preparing America's Food Safety System for the Twenty-First Century—Who is Responsible for What When it Comes to Meeting the Food Safety Challenges of the Consumer-Driven Global Economy? Food & Drug. L.J. 52, 13-30; Nestle, M., 2002. Food Politics: How the Food Industry Influences Nutrition and Health. University of California Press.

8. The White House, Office of the Press Secretary, 2011. Remarks by the President in State of Union Address.

9. U. S. General Accounting Office, 2004. Testimony Before the Subcommittee on Civil Service and Agency Organization, Committee on Government Reform, House of Representatives, Federal Food Safety and Security System: Fundamental Restructuring Is Needed to Address Fragmentation and Overlap.

10. U. S. General Accounting Office, 2004. Testimony Before the Subcommittee on Civil Service and Agency Organization, Committee on Government Reform, House of Representatives, Federal Food Safety and Security System: Fundamental Restructuring Is Needed to Address Fragmentation and Overlap.

the new requirements. Even before the new mandates kicked in, the FDA was able to oversee only a tiny fraction of the food we eat; in 2002, the agency managed to inspect just 1% of the seafood products imported into this country.[11] In these cash-strapped circumstances, the inefficiencies brought on by regulatory fragmentation are unacceptably costly.

Even where agencies charged with overlapping regulatory responsibilities for food manage to coordinate their activities to reduce duplication, the coordination itself expends precious time and resources. Coordination within the federal bureaucracy is laborious. Even if productive, it can delay the federal response to problems in the food system.[12] Moreover, just imagine the transaction costs of negotiating, mastering, and policing the 51 separate interagency agreements on regulatory authority over food; the mind fairly reels.

One might accept some inefficiency if the regulatory system were effective. But the same regulatory fragmentation that squanders time and money also almost guarantees gaps and weaknesses in the regulatory system itself. As the General Accountability Office (GAO) has repeatedly (and understatedly) observed in its many reports recommending a more unified or at least coordinated regulatory system for food, "the federal food safety system is not the product of strategic design as to how to protect the public health."[13] The failure to create a coherent, effective structure aimed at protecting public health from food-related harms manifests itself in many different ways. To take one example, the federal response to multistate outbreaks of foodborne illness is split among four different federal agencies and further divided among state and local authorities, with predictably confusing and potentially dangerous results.[14] Even the seemingly simple task of notifying the public of a suspected outbreak of foodborne illness is an exercise in jurisdictional complexity. A citizen wishing to learn about an outbreak may visit the website of the Centers for Disease and Control and Prevention (CDC)—if CDC is the lead agency investigating the outbreak. Otherwise, she needs to know whether the outbreak was caused by raw meat, raw poultry, or egg products (eggs outside the shell, not inside it); if it was, then she should repair to the USDA's website for information about the outbreak.

---

11. U. S. General Accounting Office, 2004. Testimony Before the Subcommittee on Civil Service and Agency Organization, Committee on Government Reform, House of Representatives, Federal Food Safety and Security System: Fundamental Restructuring Is Needed to Address Fragmentation and Overlap.

12. Hawthorne, F., 2005. Inside the FDA: The Business and Politics Behind the Drugs We Take and the Food We Eat. John Wiley & Sons, Inc.; U. S. General Accounting Office, 2001. Testimony Before the Subcommittee on Oversight of Government Management, Restructuring and the District of Columbia, Committee on Governmental Affairs, U.S. Senate, Food Safety and Security: Fundamental Changes Needed to Ensure Safe Food.

13. U. S. General Accounting Office, 2011. Report to Congressional Committees, GAO-11-289, Federal Food Safety Oversight: Food Safety Working Group Is a Positive First Step but Governmentwide Planning Is Needed to Address Fragmentation.

14. U. S. General Accounting Office, 2013. Report to Congressional Committees, GAO-13-283, High-Risk Series: An Update.

If the outbreak was caused by foods other than raw meat, raw poultry, or eggs outside the shell, she needs to turn to the FDA website to learn about it. Delivering timely information about an outbreak of foodborne illness—an essential part of an effective response to an outbreak—is difficult in a fractured food safety system.[15] As the National Performance Review, headed by Vice President Gore, put it, "Sometimes duplication among federal programs can make us ill—even kill us."[16]

The problems of inefficiency and ineffectiveness flowing from a disjointed food safety system are made more intractable by the fragmentation itself. Fragmentation defeats accountability; it makes it hard if not impossible to say who is responsible for a problem related to food.[17] The possibilities for deflecting blame are well illustrated by the dueling responses from the FDA and CDC in the midst of a 2002 outbreak of *Salmonella* poisoning from packaged tomatoes: the FDA said it could not recall the tomatoes because it was awaiting information from the CDC; the CDC reported that the FDA had all the information it needed.[18]

Slicing regulatory authority over food into so many institutional pieces also can defeat regulatory effectiveness because all the pieces are not equal. Although, as noted, many different federal agencies bear some responsibility for food safety and transparency, the bulk of the work falls to the FDA and to agencies within the USDA. Over 90% of the federal funding and personnel devoted to ensuring food safety and quality has historically gone to the FDA and the USDA.[19] As Michael Taylor, the current head of the FDA's office with authority over food, has put it, the United States has "two food safety systems"—the one run by the USDA, and the one run by the FDA.[20] The USDA oversees meat, poultry, and egg products, and the FDA covers essentially everything else. This bifurcation dates back over a century, to the 1906, post-*Jungle* enactment of the first two federal laws on food safety and transparency, the Pure Food and Drugs Act and the Federal Meat Inspection Act. Regulatory authority split in two directions under these laws, and it has not reconnected since. Authority over meat went to the Bureau of Animal Industry in the USDA, and authority

15. Center for Science in the Public Interest, 2001. Closing the Gaps in Our Federal Food-Safety Net.
16. Merrill, R. A., Francer, J. K., 2000. Organizing Federal Food Safety Regulation. Seton Hall L. Rev. 31, 61-173.
17. Merrill, R. A., Francer, J. K., 2000. Organizing Federal Food Safety Regulation. Seton Hall L. Rev. 31, 61-173; Taylor, M. R., 1997. Preparing America's Food Safety System for the Twenty-First Century—Who is Responsible for What When it Comes to Meeting the Food Safety Challenges of the Consumer-Driven Global Economy? Food & Drug. L.J. 52, 13-30.
18. Hawthorne, F., 2005. Inside the FDA: The Business and Politics Behind the Drugs We Take and the Food We Eat. John Wiley & Sons, Inc.
19. U. S. General Accounting Office, 1990. Report to Congressional Requesters, Food Safety and Quality: Who Does What in the Federal Government, GAO/RCED-91-19A.
20. Taylor, M. R., 1997. Preparing America's Food Safety System for the Twenty-First Century—Who is Responsible for What When it Comes to Meeting the Food Safety Challenges of the Consumer-Driven Global Economy? Food & Drug. L.J. 52, 13-30.

over other foods went to the USDA's Bureau of Chemistry (which later evolved into the FDA and shifted departments, eventually settling in the Department of Health and Human Services (HHS)).[21]

Despite receiving regulatory authority over food at the same time and in the same social and political milieu, and despite having similar legal mandates to protect the safety and transparency of the food supply, the FDA and the USDA are different in almost every other respect. They receive starkly different appropriations from Congress; the USDA's FSIS receives the vast proportion of the funding related to regulation of food, even though the FDA regulates some 80–90% of the food consumed in the United States.[22] The disparity in resources shows that while USDA inspectors continuously monitor meat and poultry passing through the nation's slaughterhouses, FDA inspectors can be expected to show up at the food processing facilities they oversee about every 5–10 years.[23] Stated in the terms of the examples of fragmentation given at the outset, this means that a facility processing a corn dog will be continuously monitored while one processing a bagel dog can wait years for a visit from a government inspector. Perhaps contributing to the disparity in resources is a gulf in political clout. The USDA sits higher than the FDA in any government organizational chart; it is a cabinet agency whose head does not answer to another agency, whereas the FDA is a subordinate part of a cabinet agency and is headed by a commissioner who ranks lower in the federal hierarchy even than the USDA's undersecretary of food safety. Less formally, the political power of the USDA relative to the FDA asserts itself in everything from dominance in developing dietary guidance to outsized influence in the White House process for reviewing agency rules.[24] The agencies also differ in their culture and mindset. Much of the USDA's job is not regulating the food industry but promoting it. The USDA's dual roles of regulation and promotion create a conflict of interest that permeates the department's work, a conflict that, formally at least, does not exist within the regulation-focused FDA.[25] None of these differences—in resources, clout, and culture—is "the product of strategic design as to how to protect the public health,"[26] and yet each of them greatly affects the quality of the food system in this country.

---

21. Nestle, M., 2002. Food Politics: How the Food Industry Influences Nutrition and Health. University of California Press; Congressional Research Service, 2014. The Federal Food Safety System: A Primer.

22. Congressional Research Service, 2014. The Federal Food Safety System: A Primer.

23. U. S. General Accounting Office, 2004. Testimony Before the Subcommittee on Civil Service and Agency Organization, Committee on Government Reform, House of Representatives, Federal Food Safety and Security System: Fundamental Restructuring Is Needed to Address Fragmentation and Overlap.

24. Nestle, M., 2003. Safe Food: Bacteria, Biotechnology, and Bioterrorism. University of California Press.

25. Institute of Medicine and National Research Council, 1998. Ensuring Safe Food: From Production to Consumption; Nestle, M., 2003. Safe Food: Bacteria, Biotechnology, and Bioterrorism. University of California Press.

26. U. S. General Accounting Office, 2011. Report to Congressional Committees, GAO-11-289, Federal Food Safety Oversight: Food Safety Working Group Is a Positive First Step but Governmentwide Planning Is Needed to Address Fragmentation.

The observations offered so far are not new. Numerous official government reports have been prepared, bemoaning this senseless division of authority and recommending consolidation of responsibility in one agency.[27] Outside groups have joined in the critiques and calls for reform.[28] Dozens of bills have been introduced in Congress to fix the problems.[29]

There is, however, one aspect of the problem of fragmentation that has largely escaped discussion. All of the critiques and proposals for reform have focused on the safety of our food rather than on the transparency of the food system. Yet safety and transparency travel together. Often, consumers' desire for transparency in the food system emerges from a desire to eat food that is wholesome for them. Moreover, transparency stands with safety as one of the principal goals of the entire regulatory system for food. Stretching back over more than a century, federal laws related to food aim to ensure both the integrity of representations made by sellers of food about their food and the safety of the food supply. The twin goals of knowledge and safety—frequently appearing as prohibitions on "misbranding" and "adulteration"—pervade numerous federal and state regulatory regimes relating to food. These goals and their attendant prohibitions run like mantras through the Food, Drug and Cosmetic Act, the federal Meat Inspection Act, the Poultry Products Inspection Act, and the Egg Products Inspection Act. Each of these laws contains the same basic injunctions against adulterated and misbranded food. In addition, the federal Lacey Act prohibits, among other things, the false labeling of seafood. The Federal Trade Commission Act prohibits "unfair or deceptive acts or practices in or affecting commerce," including such acts or practices that relate to food. The Federal Insecticide, Fungicide, and Rodenticide Act aims to ensure that pesticides are properly labeled and do not pose an unacceptable risk to humans or the environment. In sum, it is possible to connect many seemingly disparate federal laws relating to food by emphasizing the twofold goals of knowledge and safety. It is important, therefore, to consider how regulatory fragmentation affects not only the safety of our food but our knowledge of it as well.

If anything, regulatory fragmentation is more successful in defeating the goal of knowledge than it is in defeating the goal of safety. Fragmentation

27. *See* U. S. General Accounting Office, 2001. Testimony Before the Subcommittee on Oversight of Government Management, Restructuring and the District of Columbia, Committee on Governmental Affairs, U.S. Senate, Food Safety and Security: Fundamental Changes Needed to Ensure Safe Food; U. S. General Accounting Office, 2004. Testimony Before the Subcommittee on Civil Service and Agency Organization, Committee on Government Reform, House of Representatives, Federal Food Safety and Security System: Fundamental Restructuring Is Needed to Address Fragmentation and Overlap; U. S. General Accounting Office, 2011. Report to Congressional Committees, GAO-11-289, Federal Food Safety Oversight: Food Safety Working Group Is a Positive First Step but Governmentwide Planning Is Needed to Address Fragmentation; U. S. General Accounting Office, 2013. Report to Congressional Committees, GAO-13-283, High-Risk Series: An Update.

28. Center for Science in the Public Interest, 2001. Closing the Gaps in Our Federal Food-Safety Net.

29. Institute of Medicine and National Research Council, 1998. Ensuring Safe Food: From Production to Consumption.

nurtures and preserves divergent and even opposing messages from the government about the nature and quality of our food. The very same words can come to mean different things, depending on the locus of regulatory authority. The grocery store becomes—in the words of a former secretary of HHS—a modern-day "Tower of Babel," in which "consumers need to be linguists, scientists, and mind readers to understand the many labels they see."[30]

Even the processes agencies use in pursuing transparency show fragmentation and, ultimately, sows confusion. The USDA reviews and approves the labels of the products it regulates; the FDA does not. The result is a fractured process for ensuring transparency in the food marketplace. One agency has eyes on all labels before they are used, but another rests with after-the-fact measures to correct misleading labels. Yet one would be wrong to draw from this structure a conclusion that the USDA process necessarily produces more transparent results. The USDA's filter has failed to catch patently misleading labels. Recently, for example, the USDA has acknowledged that the labels it has long allowed for "enhanced" meat products—products containing added solutions of water, salt, and other ingredients—are misleading insofar as they do not convey the loss of weight in actual meat, or the gain in sodium, associated with such products, so misleading that the agency has felt bound to propose a new rule to prevent the use of the existing labels.[31] A consumer who relied on the USDA's preapproval regime as a sign of superior transparency in USDA-regulated products would be doubly misled, both by the implicit promise of government-backed transparency and by the dubious labels that pass through the USDA's filter.

Different agencies may also embrace different interpretations of the very same words. The words "naturally raised" have a precise and approved meaning when they are applied to products regulated by the USDA,[32] but the word "natural" has very little meaning when applied to products regulated by the FDA.[33] The USDA has stretched the approved meaning of the word "fresh" to cover a wide range of pretty thoroughly worked-over products, including the now-infamous "lean finely textured beef" (aka "pink slime"). The USDA has also held fast to its position that poultry remains "fresh" down to an internal temperature of 26 °F and may be called "never frozen" all the way down to an internal temperature of 0 °F. The agency will not, however, countenance the use of the term "fresh" for any product "treated with an antimicrobial substance or

30. Hilts, P. J., 2003. Protecting America's Health: The FDA, Business, and One Hundred Years of Regulation. University of North Carolina Press.

31. U. S. Department of Agriculture, 2011. Common or Usual Name for Raw Meat and Poultry Products Containing Added Solutions. Federal Register 76, 44855-44865.

32. U. S. Department of Agriculture, 2009. United States Standards for Livestock and Meat Marketing Claims, Naturally Raised Claim for Livestock and the Meat and Meat Products Derived From Such Livestock. Federal Register 74, 3541-3543.

33. Food and Drug Administration, 2014. Letter declining to define term "natural" when used on food labels.

irradiated."[34] The FDA, for its part, holds that the term "fresh" may only be applied to "raw, unprocessed foods," and may not be applied to any food product that has "been frozen or subjected to any form of thermal processing," although it embraces the use of the term for produce that has been treated with "a mild chlorine wash or mild acid wash" and for foods treated with ionizing radiation.[35] The combined consequence of these policies is that the same word, "fresh," has approved meanings that point in opposite directions, depending on the locus of regulatory authority. Examples could be multiplied. A consumer can be forgiven for being confused.

Despite the adverse consequences of regulatory fragmentation for the safety and transparency of our food supply, no appreciable progress has been made in reducing regulatory fragmentation in the United States, even as our peer countries have moved to consolidate their own regulatory systems for food.[36] Indeed, arguably, our political apparatus has chosen to respond to fragmentation not by lessening it, but by doubling down on it. Recent congressional actions relating to regulation of food have hardened rather than softened fragmentation. The 2008 farm bill, for example, added a new layer of bureaucracy to the regulation of imported catfish, duplicating existing structures and increasing expense without evident improvement.[37] The 2011 FSMA continued the existing fragmentation by giving authority to the FDA to enact new standards for food processors and produce farmers, without ameliorating the existing bifurcation of responsibilities between the FDA and the USDA.[38] In fact, FSMA arguably exacerbated fragmentation by providing that the USDA should be consulted in setting the new standards—meaning that once the standards were introduced, it was hard to know which agency had had the upper hand in developing them. In addition, President Obama's strengthening of the White House's role in reviewing regulations has also exacerbated fragmentation, by giving agencies throughout the federal government a significant but secret role in developing rules. The upshot is that once a rule has been through this wringer, one cannot know exactly which agency or individual had a decisive role in shaping it.[39] Here, too, fragmentation

---

34. U. S. Department of Agriculture, 2005. Standards and Labeling Policy Book. Available at http://www.fsis.usda.gov/OPPDE/larc/Policies/Labeling_Policy_Book_082005.pdf.
35. Food and Drug Administration, 1993. Food Labeling: Nutrient Content Claims, General Principles, Petition, Definition of Terms; Definitions of Nutrient Content Claims for the Fat, Fatty Acid, and Cholesterol Content of Food. 58 Fed. Reg. 2302-2401.
36. U. S. General Accounting Office, 2001. Testimony Before the Subcommittee on Oversight of Government Management, Restructuring and the District of Columbia, Committee on Governmental Affairs, U.S. Senate, Food Safety and Security: Fundamental Changes Needed to Ensure Safe Food.
37. U. S. General Accounting Office, 2013. Report to Congressional Committees, GAO-13-283, High-Risk Series: An Update.
38. U. S. General Accounting Office, 2011. Report to Congressional Committees, GAO-11-289, Federal Food Safety Oversight: Food Safety Working Group Is a Positive First Step but Governmentwide Planning Is Needed to Address Fragmentation.
39. Heinzerling, L., 2014. Inside EPA: A Former Insider's Reflections on the Relationship Between the Obama EPA and the Obama White House. Pace Envtl. L. Rev. 31, 325-369.

defeats accountability, yet we are moving toward fragmentation rather than away from it. As one former FDA commissioner resignedly put it, fragmentation of regulatory authority over food is "a fact of life."[40]

Perhaps we would do well, for now, to acknowledge that consolidation of food safety and transparency into one federal organization is not on the horizon, and to turn our attention to other goals. Although reorganization would make sense in so many ways, it is important to recognize that reorganization itself will not solve the most fundamental problems with our regulatory system for food. Reorganizing the bureaucracy will not bring more resources to food safety and transparency or shift those resources to more important tasks.[41] Any organization formed as a consequence of bureaucratic reshuffling would itself need to contend with difficulties of coordination; even a task assigned to one agency must often be delegated to numerous individuals or groups within the agency, with no guarantee that anyone will have a synoptic vision over the whole task.[42] Most fundamentally, bureaucratic reorganization alone would not loosen the cozy ties between the bureaucracy and industry, nor would it address the failure to understand the whole system of agriculture and food as a connected series of challenges. Better to use the time and energy that would be devoted to bureaucratic reorganization to engage in substantive renovation of our laws relating to food.

40. Hawthorne, F., 2005. Inside the FDA: The Business and Politics Behind the Drugs We Take and the Food We Eat. John Wiley & Sons, Inc.

41. Merrill, R. A., Francer, J. K., 2000. Organizing Federal Food Safety Regulation. Seton Hall L. Rev. 31, 61-173.

42. Magill, M. E., 1995. Congressional Control Over Agency Rulemaking: The Nutrition Labeling and Education Act's Hammer Provisions. Food & Drug L.J. 50, 149-190.

Chapter 11

Global Administrative Law and the FDA's Collaborative Agreements with International Regulatory Partners

Sam F. Halabi
The University of Tulsa College of Law and Scholar, O'Neill Institute for National and Global Health Law, Washington, District of Columbia, USA

Chapter Outline

The global food supply is increasingly intertwined across domestic and foreign markets in ways that challenge traditional modes of regulation. The world's largest markets, including the European Union and the United States, are representative. They are not only the largest exporters of food and beverage products, they are also the world's largest importers; so, while they are technically capable of supplying all the food their domestic populations require, specialization has resulted in some categories of food being almost entirely imported from elsewhere. In the United States, for example, 80% of seafood, 60% of fruit, and 20% of vegetables are imported, accounting for 15% of the U.S. food

1. FDA Could Strengthen Oversight of Imported Food by Improving Enforcement and Seeking Additional Authorities, GAO, May 2010.

Food and Drug Regulation in an Era of Globalized Markets. http://dx.doi.org/10.1016/B978-0-12-802311-2.00011-4

supply.[1] The volume of imported food has increased enormously over the past 20 years, from fewer than 200,000 line-entries in the early 1990s to an estimated 12 million in 2013.[2] There are now over 114,000 facilities in over 150 countries registered with the U.S. Food and Drug Administration (FDA) exporting food to the United States. Moreover, there is greater complexity in the manufacturing and labeling of foreign products (which may pass through two or more jurisdictions before reaching the United States), more foreign facilities supplying the United States, imports coming from countries with less-sophisticated regulatory systems, greater opportunities for economic fraud and, therefore, more concerns about food safety and quality.

The U.S. Food Safety Modernization Act (FSMA), covered more thoroughly by Christopher Waldrop in Chapter 13, was a sweeping and comprehensive response to the rapid and fundamental changes wrought by globalization.[3] Under the FDA's traditional import paradigm, the agency was focused on the U.S. border, using information provided through customs invoices. The FDA made decisions about the compliance status of products at the time of entry with limited time, resources, and information. As imports increased, the FDA became unable to effectively inspect incoming goods. FSMA required and/or enabled the FDA to obtain and use information from the entire life cycle of a product—where it originated, what happened to it en route to the United States—to implement more effective and comprehensive strategies. Indeed, even today, the FDA physically assesses approximately 1% of imports that cross the border. FSMA created a multilayered regulatory system that leveraged relevant information about manufacturers, importers, third-party verification entities, and foreign regulatory partners, including international organizations. It is these latter two categories of partners this chapter analyzes in the context of the broader volume.

THE FOOD SAFETY MODERNIZATION ACT AND THE PRIORITIZATION OF INTERNATIONAL REGULATORY PARTNERSHIPS

FSMA grants the FDA broad new powers to prevent food safety problems, detect and respond to food safety issues, and improve the safety of imported foods.[4] Because the law specifically aimed to update FDA authority in light of the reality of global food and food additive markets, Section 305 of FSMA called for the FDA to develop a comprehensive plan to expand the "technical, scientific, and regulatory capacity of foreign governments and their respective food industries in countries that export food to the United States."[5] Part of its

2. http://www.fda.gov/Food/GuidanceRegulation/FSMA/ucm432576.htm.
3. http://www.fda.gov/Food/GuidanceRegulation/FSMA/.
4. http://agriculture.vermont.gov/food_safety_consumer_protection/fsma.
5. US Department of Health and Human Services and US FDA, FDA's International Food Safety Capacity Building Plan v February 2013 available at http://www.fda.gov/downloads/Food/GuidanceRegulation/UCM341440.pdf.

plan for fulfilling its Section 305 obligation is actual presence: the FDA has established overseas offices in China, India, Latin America, Europe, and sub-Saharan Africa. These offices undertake a number of activities related to coordinating with foreign regulatory counterparts to strengthen capacity to assure product safety; working with regulated industries; coordinating with U.S. regulatory colleagues in-country; and, increasing the FDA's ability to perform more timely inspections. A separate aim is to incorporate the work and insights from international organizations ranging from the Food and Agriculture Organization to the World Health Organization (WHO) to the World Bank to the OECD.[6] The FSMA statute specifies one particular international organization for FDA attention with respect to its global activities, the Codex Alimentarius Commission or Codex, which Aliza Glasner discusses at greater length in Chapter 14.

The bilateral relationship between the FDA and other national regulatory authorities or between the FDA and international organizations like Codex is necessarily complex and informed not only by the FDA's mandates but by other U.S. and foreign constituencies such as, in the U.S. context, the USDA, the Department of Commerce, and the Department of State. The FDA will therefore adopt a range of strategies to accomplish its objectives within complex national and international frameworks. Systems recognition, for example, is a process the FDA uses to determine whether a food safety system provides a similar, though not identical, set of public health protections as the U.S. food safety system. For a food safety system to be recognized, the FDA considers whether foods in that country are produced under standards that offer the same level of public health protection as U.S. standards.

The FDA's mission includes the mandate to participate through appropriate processes with representatives of other countries to reduce the burden of regulation, harmonize regulatory requirements, and achieve appropriate reciprocal arrangements. In 2011, the FDA released its report, *Pathway to Global Product Safety and Quality*.[7] This document outlined both the near-term changes that would affect the U.S. supply of drugs, food, and medical devices as well as the strategies the FDA would pursue to address those changes. *Pathway* commits the FDA to building global coalitions of regulators dedicated to building and strengthening the product safety net around the world. Those coalitions would, in turn, focus on information sharing and intelligence gathering aimed at product integrity and food safety. This chapter analyzes current agreements and international arrangements, of which there are over 100 between the FDA and regulatory partners, not only to understand how the principal U.S. food safety agency approaches international collaboration, but also how collaborative arrangements may be specific to certain national regulatory relationships, products, aspects of products, or information that will influence the scope and

---

6. *Id.* at 3.
7. http://www.fda.gov/downloads/AboutFDA/CentersOffices/OfficeofGlobalRegulatoryOperationsandPolicy/GlobalProductPathway/UCM262528.pdf.

dimensions of the global regulatory coalitions envisioned in *Pathway*. It is worth noting that while the agreements analyzed in this chapter are focused on food safety and cooperation over food safety-related measures, parallel agreements are in place for pharmaceuticals and medical devices.

In undertaking this analysis, this chapter pursues two objectives. First, within the burgeoning literature on "global administrative law," the FDA's approaches to international agreements suggest that the kinds of activities undertaken in the "global administrative space" may not be so straightforwardly "administrative" as scholars working within the literature now suggest, so that the phrase may be, so far as its stated subject matter, underinclusive. Indeed, the FDA has entered into international agreements, I argue, not only for their specific administrative or regulatory purpose, but also to encourage regulatory partners with complex or opaque regulatory systems to bring greater clarity and understanding to those systems, especially as they become larger exporters of regulated products to the United States. Second, identifying, analyzing, and classifying the FDA's international agreements provides models, templates, and lessons for national regulatory authorities seeking to build their partnerships not only with the FDA but closer regional or international regulatory partners that share the FDA's national consumer protection mandate.

INTERNATIONAL AGREEMENTS: TYPOLOGIES AND OBJECTIVES

In 2005, Benedict Kingsbury, Nico Krisch, and Richard Steward published their foundational article, *The Emergence of Global Administrative Law*, in which they argued that a discrete if amorphous body of law had emerged comprised of transgovernmental regulation and administration "designed to address the consequences of globalized interdependence in such fields as security, the conditions on development and financial assistance to developing countries, environmental protection, banking and financial regulation, law enforcement, telecommunications, trade in products and services, intellectual property, labor standards, and cross-border movements of populations, including refugees."[8] While food and drug national regulatory authorities played little if any role in their discussion, Kingsbury, Krish, and Steward identified precisely the kind of work the FDA undertakes with respect to its global regulatory partnerships: administration based on collective action by transnational networks of cooperative arrangements between national regulatory officials. Yet their assessment of what those networks do, and why, strongly implied that regulators participating in collaborative networks were driven primarily by their "regulatory mission" toward a relatively discrete set of objectives like internal administrative accountability, protection of rights, or the promotion

8. Benedict Kingsbury, Nico Krisch, Richard Steward, The Emergence of Global Administrative Law, 68 Law and Contemporary Problems, 15 (2005).

of democracy. The FDA's regulatory partnerships, by contrast, not only take a broader view of what their consumer protection mandate entails, but also the methods by which any given aspect of that mission might be advanced through bilateral or multilateral agreements.

The FDA has, through its Office of Global Regulatory Operations and Policy (which reports directly to the commissioner's office), long engaged foreign counterparts in exploring areas of mutual interest and forging common regulatory standards and review processes that expand access to safe drugs, medical devices, and food for both U.S. and foreign populations. For human drugs, the FDA has since 1990 worked with the International Conference on Harmonisation of Technical Requirements for Registration of Pharmaceutical Products for Human Use (ICH). Through the ICH process, American, European, and Japanese drug regulatory authorities and pharmaceutical industries meet to discuss and formulate recommendations toward achieving greater harmonization in the interpretation and application of technical guidelines and requirements for pharmaceutical product registration, reducing or obviating duplication of testing carried out during the research and development of new human medicines. Those recommendations in turn influence harmonization efforts globally. For inspection of pharmaceutical products, FDA has joined the Pharmaceutical Inspection Cooperation/Scheme (PIC/S), an informal organization of drug manufacturing inspectorates from 39 countries that share pharmaceutical facility inspection information. For veterinary drugs, the International Cooperation on Harmonisation of Technical Requirements for Registration of Veterinary Medicinal Products, or VICH, endeavors to harmonize technical requirements for veterinary product registration. For medical devices, FDA cooperates with the International Medical Device Regulators Forum, or IMDRF, a voluntary group of medical device regulators who aim to accelerate international medical device regulatory harmonization and convergence. IMDRF was born in October 2011, when representatives from the medical device regulatory authorities of Australia, Brazil, Canada, China, European Union, Japan, and the United States, as well as WHO met in Ottawa to address the progress of preexisting medical device harmonization initiatives.

Participation in harmonization initiatives, however, is quite different than negotiating and reaching common understandings with foreign counterparts that may be reduced to writing, an act which, whatever the technical legal consequence, exercises real influence on mutual expectations and conduct over critical consumer goods areas. The discussion below outlines the key agreements the FDA pursues and implements to further its consumer protection mandate, which range from nonspecific cooperative commitments as to intelligence sharing to product-specific and risk-specific memoranda of understanding or exchanges of letters meant to address priority consumer safety issues within the context of bilateral or multilateral trade and economic relationships to strongly worded "agreements" that contain treaty-like language and expectations.

Confidentiality Agreements

The FDA has over 120 international agreements in place with about 30 countries and multilateral partners. Of these, nearly half are confidentiality commitments that allow the FDA to share information, protect information disclosed by regulatory partners, and take measures to ensure that any disclosure of information provided by a regulatory partner is done so according to law and with minimal disturbance to the regulatory partner's mission. FDA and the Controlled Substances and Tobacco Directorate of the Healthy Environments and Consumer Safety Branch of Health Canada, for example, have entered into a detailed agreement governing the sharing of information related to tobacco control.[9] While confidentiality commitments between the FDA and regulatory partners begin with a standard one-page template, there is significant variation between the terms that the FDA is able to negotiate and memorialize. Confidentiality commitments between the FDA and the European Food Safety Agency on the one hand, and the FDA and Mexico's Secretaria de Agricultura, Ganaderia, Desarrollo Rural, Pesca y Alimentacion (SAGARPA), on the other, reflect priority concerns each side maintains with respect to applicable laws, ongoing scientific cooperative efforts, and other relationship-specific factors.

Statements of Cooperation or Intent

If confidentiality commitments represent the narrowest form of international regulatory cooperation the FDA undertakes (which may explain the large number the FDA has in place), statements of cooperation represent a next step toward more extensive cooperative activities. Statements of cooperation or intent outline general contours of product-specific regulatory possibilities with partners with robust cooperative histories or foreshadow greater cooperation with regulatory partners with whom cooperative activities are not well established. In 2005, for example, the FDA entered a statement of intent with Mexican agencies SENASICA and COFEPRIS to explore a work plan that would "achieve mutual confidence" in the produce safety systems for fresh and minimally processed product. In 2012, the FDA entered into a statement of cooperation with the Brazilian health regulatory authority ANVISA regarding "cooperation to enhance activities of mutual interest." The statement reads very much like two parties who have infrequently dealt with one another and, even when so, have not endeavored to obtain a full understanding of the other's regulatory approaches. The statement of cooperation outlines five categories of information exchange, collaboration, mutual investigation, and shared research interests that might form the basis of more detailed regulatory cooperation in the future.[10]

---

9. http://www.fda.gov/InternationalPrograms/Agreements/ConfidentialityCommitments/ucm290778.htm.

10. http://www.fda.gov/InternationalPrograms/Agreements/MemorandaofUnderstanding/ucm330164.htm.

Exchanges of Letters

Where the FDA is unable to reach mutually agreed terms, bilateral terms are generally expressed as "positions" with tacit understanding as to differences held by regulatory partners, or where special circumstances otherwise obtain, the FDA enters into "exchanges of letters," which articulate the respective positions of the parties with the aim of reducing, if not eliminating, barriers to regulatory cooperation. For example, puffer fish contain a potentially lethal neurotoxin, tetrodotoxin, in their internal organs and even skin but are regarded as a delicacy in China, Korea, and Japan where specially trained chefs know which parts of the fish to safely prepare and in what quantity. Puffer fish have long been on the FDA's list of automatic detention products precisely because there is no way to approve an incoming "lot" of the fish; any one fish may contain a lethal amount of tetrodotoxin. In 1988, the FDA and the Japanese Ministry of Health and Welfare entered into an exchange of letters under which exporters following a specific protocol would address the FDA's concerns and thus encounter fewer problems at the border.[11] In 1995, the FDA entered into an exchange of letters with the New Zealand Ministry of Agriculture and Fisheries that committed the FDA to a 99% "may proceed rate" border approval upon certain certifications from the New Zealand agency, although the final letter in the exchange made clear that each party retained certain "understandings" with respect to the approval program.

Memoranda of Understanding

The FDA also uses memoranda of understanding and other cooperative arrangements, which do not have a standard form like confidentiality commitments and take a variety of forms that suit specific regulatory or product factors. The purpose of these agreements is to memorialize intended cooperation with foreign partners with the goal of improved safety and quality of regulated goods and components imported into the United States. They help to facilitate a greater understanding of foreign regulatory systems, and even when they do not lead to a perfectly harmonized system, they facilitate FDA cooperation with and mutual reliance on their foreign counterparts.

The 2012 Memorandum of Understanding (MOU) between the FDA and the Mexican National Agro-Alimentary Health, Safety and Quality Service SENASICA Concerning the Entry of Mexican Cantaloupe into the United States is a good example of a product-specific agreement. Cantaloupe and other melons present special food safety risks. Netted melons like cantaloupe grow on the ground and can come in contact with pathogens in noncomposted fertilizer or through handling. Unlike other fruits, cantaloupe are not acidic and readily

11. http://www.fda.gov/InternationalPrograms/Agreements/MemorandaofUnderstanding/ucm107601.htm.

support the growth of pathogens once they are sliced open. Outbreaks of illness linked to melons contaminated with salmonella occur each year.[12] Foodborne illness outbreaks from cantaloupe have been traced back to wash water, shipping ice, and even contact with contaminated meat.[13] For these reasons, the FDA categorizes cut melon as a potentially hazardous food and it therefore receives substantial regulatory attention from the FDA, especially at the border. Although imports of Mexican cantaloupe have decreased from the 1970s and 1980s when Mexican farmers supplied 90% or more of the fruit to U.S. markets, Mexican cantaloupe still account for a substantial portion of the fruit consumed in the United States.[14] The MOU facilitates the implementation of a system that increases the likelihood that cantaloupe from Mexico that will be imported into the United States complies with U.S. law, and facilitates a risk-based system to help ensure that the cantaloupe is safe for consumption. In conjunction with this agreement, the FDA also intends to accept the SENASICA Official Certificate and supporting documentation to help implement Systems for the Reduction of Contamination Risks (SRRC) during primary production of cantaloupe. Other examples of product-specific agreements include several MOUs regulating dry milk products, shellfish safety, milk proteins, and fish. MOU relationships can be much broader and context specific. The 1998 Record of Understanding Between the Governments of the United States and Canada Regarding Areas of Agricultural Trade, for example, included measures to be taken by a wide range of U.S. and Canadian agencies, with specific aspects of veterinary drugs and horticultural produce committed to FDA action. The 2013 MOU between the FDA and China's Center for Food Safety Risk Assessment outlines nine discrete categories of "projects of mutual interest" the parties intend to undertake collaboratively.

Systems Recognition

Systems recognition arrangements differ from MOUs and exchanges of letters in that they represent broader, comprehensive assessments of regulatory objectives and capability and offer mutual recognition based on comparable levels of protection of public health, even if specific protocols or processes may differ. Indeed, the concept of systems recognition is a relatively new one for the FDA, although it was based on previous assessment templates, the International Comparability Assessment Tool created as an objective framework for assessing the robustness of trading partners' overall food safety systems, in turn based on the Manufactured Food Regulatory Program Standard (MFRPS), which is an assessment tool that the FDA utilizes to provide a uniform foundation for the design and management of state programs responsible for the regulatory oversight

12. http://www.foodsafety.wisc.edu/assets/pdf_Files/Safety_of_Cantaloupe.pdf.

13. 2005 Study International Journal of Food Microbiology.

14. USDA, Effects of the North American Free Trade Agreement on Agriculture and the Rural Economy 123 (2002).

of food plants. The sheer cost and extent of the systems recognition process underscores Commissioner Hamburg's view that formal bilateral agreements are costly relative to the benefits they impart. The in-country review trip of the FDA to New Zealand took place in September 2010. Five teams participated in the in-country review, each focusing on a specific area of the assessment. The in-country review of New Zealand's food safety system involved verification of the implementation of the New Zealand food safety system through documentation review of the program's records, as well as on-site visits. In conjunction with the comparability review, site visits focused on the observation of inspection and systems audit programs at seafood and dairy processing firms and a review of the New Zealand shellfish sanitation program for growing area classification and related controls.

In 2012, after the FDA's extensive review of New Zealand's food safety control system, the FDA and New Zealand's Ministry for Primary Industries (MPI) entered into a Food Safety Systems Recognition Arrangement that acknowledged that the FDA and MPI have comparable food safety regulatory systems that lead to a similar level of food safety assurance. Reaching that conclusion and codifying it required lengthy mutual assessment of the operation of the food control systems in effect in each country as well as more in-depth reviews of the regulatory systems in place for commodities of higher regulatory interest like dairy products and seafood (including shellfish). The systems recognition agreement indicates an intention for the two agencies to accomplish two tasks: first, that they will work cooperatively when food shipped between them has safety issues, and second, that they will work cooperatively to take any appropriate risk management interventions needed to ensure food safety while minimizing trade disruptions. For example, with New Zealand, it may use some of the data MPI generates to make decisions about New Zealand product imports. MPI data allow the FDA to prioritize resources dedicated to foreign facility inspections, import field exams, and import sampling.[15] New Zealand was the first country to obtain systems recognition status with the FDA and the FDA has disseminated a preliminary guide for other regulatory partners who are interested in systems recognition arrangements.

Agreements Within Broader Bilateral Cooperative Partnerships

On February 4, 2011, Canadian Prime Minister Stephen Harper and U.S. President Barack Obama announced the creation of the Canada-United States Regulatory Cooperation Council (RCC) to increase regulatory transparency and coordination between the two countries. In general, Canadian and U.S. delegates to the RCC undertake efforts to better align the regulatory environment between

15. http://blogs.fda.gov/fdavoice/index.php/2013/09/fda-systems-recognition-ensuring-imported-foods-are-safe/.

Canada and the United States through enhanced technical collaboration, mutual recognition of standards, and joint work sharing, in an effort to address root causes of regulatory misalignments, develop lasting solutions, and avoid future misalignments from developing. The initial Joint Action Plan seeks to foster new approaches to regulatory alignment and serve as a template for future efforts at Canada-U.S. coordination. This work plan establishes collaboration between the FDA and the Canadian Food Inspection Agency (CFIA), Health Canada, to promote regulatory alignment and mutual reliance in areas such as product review and the inspection of facilities. This agreement will also enhance collaboration on enforcement and compliance by increasing mutual reliance on each partner's inspections in order to reduce unnecessary duplicate inspections. The CFIA and the FDA will assess each other's food safety systems (excluding meat, poultry, and eggs products) and establish a food-safety systems recognition arrangement based on the results. The exchange of food safety information will be increased to help both countries make informed risk-based regulatory decisions. For example, the CFIA and the FDA have compared Canadian and U.S. accreditation criteria for food testing laboratories and analyzed the similarities between them. This analysis examined changes to food safety laboratory accreditation requirements that are possible as part of broader food safety legislation modernization efforts in both countries and that are expected to result in more aligned requirements between Canada and the United States. Moving forward, Canadian and U.S. food safety regulators will endeavor to jointly develop criteria, processes, and tools to facilitate mutual reliance on food testing results under the direction of an existing Canada-FDA Joint Committee on Food Safety.[16] As the FDA and the CFIA implement their modernized food safety rules and regulations, they will work together to ensure that their systems deliver comparable public health outcomes and reduce unnecessary duplication wherever possible. In short, the FDA-CFIA initiative within the RCC plays an analogous role to systems recognition, which is a stand-alone arrangement with New Zealand.

Agreements or Quasi-Treaties

Rarely, FDA will enter into agreements that, while they are not strictly speaking treaties within the meaning of Article II of the U.S. Constitution, nor executive agreements as they are typically used in U.S. diplomacy, wear all the insignia of binding agreements. The 2007 agreement between FDA (and, more broadly, the U.S. Department of Health and Human Services) and China's General Administration of Quality Supervision, Inspection and Quarantine on the Safety of Food and Feed, provides a relatively lengthy provision defining terms, mandating specific action from each party, including production of "copies of, and other relevant information concerning, their respective laws and regulations

16. http://www.whitehouse.gov/sites/default/files/omb/oira/irc/us-canada-rcc-joint-forward-plan.pdf.

relating to food and feed safety," and specifying points of contact for covered cooperative activities and undertakings.[17] Unlike most FDA agreements, the agreement with AQSIQ includes extensive annexes, expiration and termination provisions, and specific commitments within each aspect of the agreement.

CONCLUSION

As demonstrated by these examples, FDA cooperation with foreign partners can take many forms, challenging the unitary functions now commonly attributed to regulators participating in transgovernmental networks. The FDA's agreements may be at once inconsistent with the letter and expectations of international trade law yet also reducing the likelihood that regulatory measures might be challenged. Similarly, the "softness" or "hardness" of the agreement may aim not at core regulatory objectives of the FDA but rather in establishing a specific kind of regulatory relationship or encouraging the dissemination of official or at least more reliable information from regulatory partners. As globalizing forces increasingly intertwine consumer protection, trade, and financial regulators, the complexity of international networks and related international institutions will need to develop methodologies to assess which role any given participant is playing. From a purely pragmatic perspective, the FDA has developed a sophisticated and diverse set of agreements to manage complex relationships that may serve as useful models as national regulatory authorities across the world see the efficiencies that may be realized through use of confidentiality commitments, statements of cooperation, exchanges of letters, memoranda of understanding, or even stronger agreements and treaties.

17. http://www.fda.gov/InternationalPrograms/Agreements/MemorandaofUnderstanding/ucm107557.htm.

Chapter 12

Effective Regulation of Food Additives

The Pew Charitable Trusts
The Pew Charitable Trusts, Philadelphia, Pennsylvania, USA

Chapter Outline

Chemical additives are a fundamental part of our modern food supply. They provide flavor, enhance taste, improve appearance, add nutrient value, and prevent spoilage.[1] They are also used in food preparation and packaging materials. Their use enables consumers to have access to the food they want when they want it. But additives can also be controversial.[2]

---

* This chapter originally appeared as a report from The Pew Charitable Trusts entitled *Fixing the Oversight of Chemicals Added to Our Food: Findings and Recommendations of Pew's Assessment of the U.S. Food Additives Program.*

1. Thomas G. Neltner, Neesha R. Kulkarni, Heather M. Alger, Maricel V. Maffini, Erin D. Bongard, Neal D. Fortin, and Erik D. Olson, "Navigating the U.S. Food Additive Regulatory Program," Comprehensive Reviews in Food Science and Food Safety 10 (2011): 342-343. Accessed August 2, 2013. doi: 10.1111/j.1541-4337.2011.00166.x.

2. Maricel V. Maffini, Heather M. Alger, Erik D. Olson, and Thomas G. Neltner, "Looking Back to Look Forward: A Review of FDA's Food Additives Safety Assessment and Recommendations for Modernizing its Program," Comprehensive Reviews in Food Science and Food Safety 12 (2013): 439-440, 449. Accessed August 2, 2013. doi: 10.1111/1541-4337.12020; Thomas G. Neltner, Heather M. Alger, James T. O'Reilly, Sheldon Krimsky, Lisa A. Bero, Maricel V. Maffini, "Conflicts of Interest in Approvals of Additives to Food Determined to be Generally Recognized as Safe: Out of Balance," Journal of American Medical Association–Internal Medicine, E2, Accessed August 13, 2013 (in press). doi: 10.1001/jamaintermed.2013.10559.

Food and Drug Regulation in an Era of Globalized Markets. http://dx.doi.org/10.1016/B978-0-12-802311-2.00012-6

The Food Additives Amendment of 1958, passed by Congress and signed by President Dwight D. Eisenhower, designed a system to handle 800 additives—far fewer than the 10,000 allowed today.[3] Most new chemical additives were supposed to go through a formal process of agency review that would be triggered when a company submitted a food additive petition.[4] The petition process requires the FDA to notify the public; provide an opportunity for comment; and, if the agency deems the chemical's intended use to be safe, issue a regulation allowing the use. The law provided an exemption for common food ingredients, such as oil and vinegar, when their use is "generally recognized as safe," or GRAS. When a chemical's use is designated as GRAS, the formal public notice and comment rule making process is not required.[5]

The GRAS exemption on its face sounds very straightforward. For an additive to qualify, its use must be safe, and that safety must be generally recognized by scientists knowledgeable about the safety of substances added to food.[6] But the food industry and the FDA over the years have come to interpret the law as allowing additive manufacturers to determine that a chemical's use is safe without notifying the agency.

Initially, the GRAS exemption was used by the flavor industry for its products and by manufacturers of additives that were in common use before 1958.[7] Many of the 10,000 additives were authorized by the FDA or the food industry by the end of the 1960s.[8]

In 1969, because of safety concerns, President Richard M. Nixon ordered the agency to reassess the safety of additives and review all its prior decisions.[9] In 1982, an FDA-convened expert panel called the Select Committee on GRAS Substances completed its review of approximately 400 GRAS substances. The panel also laid out a rigorous approach to evaluating a chemical's safety and suggested improvements to the agency's process.[10]

In the 1980s, the agency laid out a comprehensive Priority-Based Assessment of Food Additives program to reevaluate its previous decisions. This program set priorities but did not simplify the process the agency followed to reverse additive decisions due to new science or other concerns. This formal rule-making process provided manufacturers with a right to an administrative hearing in most circumstances. As a result, only a handful of additives were ever reassessed, and citizen petitions calling for restriction have often not been resolved.[11] With

3. Neltner, Navigating, 367.
4. Degnan, FDA's Creative Application, 25.
5. 21 USC §321.
6. Neltner, Navigating, 351.
7. Neltner, Navigating, 347.
8. Degnan, FDA's Creative Application, 19.
9. Maffini, Looking Back, 439-440.
10. Maffini, Looking Back, 439-440.
11. Government Accountability Office, "Food Safety: FDA Should Strengthen Its Oversight of Food Ingredients Determined to Be Generally Recognized as Safe (GRAS)," 27. Accessed August 5, 2003, http://www.gao.gov/products/GAO-10-246.

limited resources and an increasingly complicated rule-making process, the FDA had an overwhelming backlog of unresolved reviews by the early 1990s. In response, in 1997 it began accepting voluntary notifications from additive manufacturers claiming that their chemicals are GRAS, which the agency would informally review.[12] The goal of companies submitting these notices is to persuade the agency to issue "no question letters." In some cases, these are subsequently cited as evidence of FDA clearance, although the agency maintains that the letters are informal and do not constitute approval. Today, virtually all new chemical additives added directly to food go through the GRAS exemption: This loophole has effectively swallowed the law.[13]

In 2010, the chairs of two congressional panels with jurisdiction over food safety, Senator Tom Harkin (D-IA) and Representative Rosa DeLauro (D-CT), asked the Government Accountability Office to scrutinize the FDA's GRAS program. Later that year, the nonpartisan investigative arm of Congress concluded that:

- "FDA's oversight process does not help ensure the safety of all new GRAS determinations."
- "FDA is not systematically ensuring the continued safety of current GRAS substances."[14]

GAO made a series of six recommendations, but the FDA has made significant progress on only one of them, by issuing draft guidance on nanoengineered particles in 2012.

In 2011, Congress passed the FDA Food Safety Modernization Act to address long-standing concerns, primarily involving pathogens. The law directs the agency to promulgate a series of regulations to prevent unsafe food from entering the market and gives it more authority to act if problems are found. One of these regulations, requiring food manufacturers to conduct a formal hazards analysis and have written risk-based preventive controls to minimize these hazards, was supposed to be finalized by July 2012. These rules have yet to be finalized, however, with only a proposal issued for comment in January 2013. When the regulations are completed and in effect, industry will have to evaluate in accordance with the rules its processes to ensure that no unapproved food additives are used in its products.

Overall, Pew estimates that more than 10,000 chemicals are permitted to be used in human food, about half as direct additives and the balance in packaging or other food-contact materials.[15] The number is not in itself a problem. Rather, it is a reflection of the diversity of the food supply and the ingenuity of industry. It also hints at the challenges facing the FDA in ensuring that these chemical additives are safe.

Pew determined that the FDA has not reviewed the safety of about 3000 of the 10,000 additives allowed in food.[16] An estimated 1000 of these 3000 are

12. Neltner, Navigating, 360.
13. Neltner, Navigating, 361.
14. GAO, FDA Should Strengthen, 20.
15. Neltner, Navigating, 355.
16. Neltner, Navigating, 355.

self-affirmed as GRAS by additive manufacturers without notice to or review by the agency, with the balance affirmed as GRAS flavors by an expert panel convened by the flavor industry trade association.[17] The FDA monitors but does not review these flavor industry decisions.

PEW'S APPROACH

In 2010, Pew launched its food additives project to:

- Evaluate the federal regulatory program designed to ensure that chemicals added to food are safe, including an examination of how it responds to advances in scientific understanding of chemical safety and changing uses of chemicals in food.
- Identify and assess viable, evidence-based, expert-vetted policy solutions if flaws are found in the regulatory program.
- Educate policy makers and key stakeholders, including industry, public interest groups, and medical associations, about the project's findings.[18]

To accomplish these objectives, Pew hired a team of scientists and lawyers to conduct the review, assembled expert advisers to guide our work, arranged for the American Academy of Pediatrics to provide critical review, committed to a transparent process that engaged stakeholders, and published our findings in peer-reviewed scientific journals.

Pew spent more than 2 years conducting research. As part of this research, we wrote an article that provided a comprehensive analysis of the U.S. food additive regulatory program, including its history and recent trends, which was published in a peer-reviewed journal of the Institute of Food Technologists, the professional society of food scientists. Recognizing the global marketplace for additives, we commissioned an industry consulting firm and a professor of international food science to compare food additive laws in other developed countries. We also held two workshops focused on the science used to make safety decisions. These workshops were cosponsored by the journal *Nature* and the Institute of Food Technologists, with the FDA participating in them and assisting in their design. More than 70 experts from industry, academia, government agencies, and public interest organizations participated in each event. We published the proceedings of each of these workshops in the Institute of Food Technologists' journal.[19]

17. Neltner, Navigating, 355.
18. Food Additives Project, The Pew Charitable Trusts. Accessed August 5, 2013, http://www.pewtrusts.org/en/archived-projects/food-additives-project.
19. Maricel V. Maffini, Heather M. Alger, Erin D. Bongard, Thomas G. Neltner, "Enhancing FDA's Evaluation of Science to Ensure Chemicals Added to Human Food Are Safe: Workshop Proceedings," Comprehensive Reviews in Food Science and Food Safety 10 (2011): 321. Accessed August 2, 2013. doi: 10.1111/j.1541-4337.2011.00165.x; Heather M. Alger, Maricel V. Maffini, Neesha R. Kulkarni, Erin D. Bongard, Thomas Neltner, "Perspectives on How FDA Assesses Exposure to Food Additives When Evaluating Their Safety: Workshop Proceedings." Comprehensive Reviews in Food Science and Food Safety 12 (2013): 90. Accessed August 2, 2013. doi: 10.1111/j.1541-4337.2012.00216.x.

In 2012, Pew shifted from research to analysis of policy options. We held a workshop with multiple stakeholders, again in collaboration with *Nature* and the Institute of Food Technologists and with continued participation from the FDA, to develop and critique potential recommendations proposed by participants. We also supported further research into three critical issues: nanoparticles, endocrine disruptors (substances that impair our hormones), and use of cell-based tests to identify chemical hazards (known as Tox21). To support this research, we held an additional workshop, supported three others, and funded an industry-affiliated think tank that convened a multidisciplinary team to evaluate methods of measuring exposure to nanoengineered particles in food.

With this foundation in place, in 2013 we identified three core issues and published our analysis in three peer- reviewed journal articles. These issues are:

- Conflicts of interest that arise when an additive's manufacturer selects the scientist who makes the GRAS safety decision.
- Data gaps in toxicity testing for additives that have been previously approved by the FDA.
- The FDA's reliance on outdated science to assess the safety of chemical additives.

Finally, we hosted a fifth workshop to address potential conflicts of interest in more depth and submitted proposed guidance to the FDA to help it resolve the issue and implement one of the Government Accountability Office's recommendations on this problem.

PEW'S FINDINGS

Our analysis focused on the overall regulatory system that is expected to ensure the safety of more than 10,000 chemical additives, rather than on concerns raised about specific substances. We evaluated the FDA's ability to fulfill the mission of its food additive regulatory program to protect public health from chemicals intentionally added to food or food packaging. We did not evaluate whether specific chemicals or groups of substances, such as salt, trans fat, caffeine, bisphenol A (which is used to line the inside of cans), or artificial colors or flavorings, cause actual harm to the public. We also did not consider contaminants found in food from natural sources or because of pollution, because those are not intentionally added and are regulated under a different set of health and safety standards.

Our research found that the FDA regulatory system is plagued with systemic problems that prevent the agency from ensuring the use of food additives is safe. If one of these chemicals were causing health problems short of immediate serious injury it is unlikely that the FDA would detect the problems unless the food industry alerted it. This is particularly true if the health consequences of ingesting the additive take years or decades to become manifest after the food is eaten.[20] If the agency did identify a problem, it would still face challenges proving harm. Proof of harm was not the safety standard laid out by Congress in 1958. Under the law, a chemical may be used in food if competent scientists are reasonably certain that the use will cause no harm over a lifetime. In short, the question is whether it will cause no harm, rather than whether harm can be proven.

Under the law, the FDA is supposed to make a determination only after it has considered the cumulative effects from similar chemicals and has information ensuring an adequate margin of safety.[21] But it is essentially impossible for the agency to connect an additive to a health problem when it has

- Not been notified about an estimated 1000 chemicals currently allowed in food.[22]
- Not been informed of actual usage for all chemicals.[23]
- Not been alerted to studies that suggest previously unknown potential health effects.[24]

In contrast, Congress in 1976 gave the U.S. Environmental Protection Agency greater, albeit still limited, authority to get more of the information it needed to make safety decisions for virtually all chemicals that are used in consumer products not regulated by the FDA. In 1996, Congress gave the EPA additional authority to protect the public from pesticide residues that may be in food.[25]

Despite these fundamental limitations and ongoing resource challenges, the FDA has

- Pursued implementation of its many responsibilities under the FDA Food Safety Modernization Act, especially the hazard analysis and preventive control requirements that should improve food additive safety.[26]
- Partnered with the EPA and the National Institutes of Health on the federal Tox21 project, which set the stage to move chemical safety work into the twenty-first century.
- Launched the Advancing Regulatory Science Initiative, in response to concerns raised by its Science Board, to help the FDA develop tools, standards, and approaches to assess the products it regulates.
- Secured voluntary industry commitments to restrict use of existing additives when concerns have arisen.[27]

---

20. Neltner, Conflicts of Interest (in press), E2.
21. Neltner, Navigating, 351.
22. Neltner, Navigating, 354.
23. Alger, Perspectives, 118.
24. Maffini, Enhancing, 334.
25. EPA, Implementation of Requirements under the Food Quality Protection Act (FQPA). Accessed August 5, 2013, http://www.epa.gov/pesticides/regulating/laws/fqpa/fqpa_implementation.htm.
26. Neltner, Navigating, 357. See also FDA, "FDA Actions to Date." Accessed August 22, 2013, http://www.fda.gov/Food/GuidanceRegulation/FSMA/ucm239907.htm
27. Maffini, Looking Back, 444.

- Developed methods to estimate exposure to chemicals that are proposed for use in food.[28]
- Supported its food additive scientists, helping some of them to become recognized experts among their peers in the international community.[29]

In summary, Pew's analysis documented that while the FDA has made efforts to improve oversight, a number of serious problems with the food additive regulatory system have led us to conclude that the Food Additives Amendment of 1958 is not working as Congress intended. Specifically, we found:

- *Conflicts of interest.* Food manufacturers make GRAS safety decisions without the FDA's knowledge despite conflicts of interest among those making the determinations. The GRAS loophole as currently used is inconsistent with Congress's plan and the practices of other developed countries.
- *Lack of information.* The FDA lacks even basic information needed to assess the safety of thousands of chemicals that have been cleared for use in food. As a result, the agency reevaluates the safety of only a relative handful of existing additives.
- *Outdated science.* The FDA uses outdated science to evaluate additive safety. It relies on a process that does not ensure independent scientific input and is often not transparent, particularly for food-contact substances.
- *Missed safety deadlines.* The agency has fallen behind the FDA Food Safety Modernization Act rule-making deadlines. Until those rules are in place, the agency has limited means to identify compliance problems.

Conflicts of Interest

Pew estimates that food manufacturers have designated 1000 chemicals as "Generally Recognized as Safe" without the FDA's knowledge. This total does not include more than 100 chemicals that were reviewed by the agency for the first time through its voluntary GRAS notification program. It also does not include industry GRAS safety decisions involving existing additives that expand the allowed uses to different foods, allow increased concentrations in food, or accommodate different manufacturing processes or purities.

As noted above, Congress required the food industry to use a petition process to obtain FDA review and approval for safety decisions on additives. Only if there was "consensus" among scientists that the use was generally recognized as safe could a chemical avoid the petition route and be declared GRAS. From 2003 to 2012, however, only 23 food additive petitions were submitted. A handful of these were for changes to existing agency approvals, not new chemicals.

28. Alger, Perspectives, 117.
29. JECFA, Joint FAO/WHO Expert Committee on Food Additives (JECFA) Roster of Toxicological and Epidemiological Experts to Serve as Members or Temporary Advisers 2011-2015. Accessed August 5, 2013, http://www.who.int/foodsafety/chem/jecfa/Roster_JECFA_2011.pdf.

In contrast, during this same time, food manufacturers submitted 332 GRAS notifications for more than 100 chemicals to the FDA seeking letters from the agency saying it had "no questions" about the safety decision. This total does not include the safety decisions made by manufacturers that they chose not to provide to the agency.

Of the 451 GRAS notifications voluntarily submitted to the FDA for review from 1997 to 2012, Pew found that financial conflicts of interest in these decisions are ubiquitous. Our findings relied on a conflict-of-interest framework developed by the Institute of Medicine in 2009. There is no basis to assume that the decisions withheld from agency review are any better. This lack of independent review raises concerns about the integrity of the process and the safety of the food supply, particularly when the manufacturer does not notify the FDA.

The GRAS exemption has implications beyond the safety of a specific additive. It hinders the agency's efforts to modernize its science, because if the FDA asks tougher questions, then firms may be less likely to voluntarily inform it.[30] It also raises the issue of whether an additive can be recognized as safe while its identity and uses are kept secret.

Our analysis confirms the GAO's conclusion that "FDA's oversight process does not help ensure the safety of all new GRAS determinations."[31] In an increasingly global marketplace where additives and food are imported into the United States, this loophole presents a situation that undermines public confidence in the safety of food and raises questions about the FDA's ability to ensure the protection of public health. Until conflicts of interest are minimized and safety decisions are subject to FDA review, the safety of food additives will largely depend on the integrity and competence of food manufacturers.

Lack of Information

Our investigation found that most additives are not tested for safety in accordance with the FDA's limited testing recommendations. Agency guidelines, for example, say that chemicals intentionally added to food should be fed to laboratory animals to identify potential harmful effects, but we found that in the majority of cases chemicals directly added to food did not undergo this very basic test. The data gaps are not significantly better for the chemicals reviewed by the agency under its voluntary notification program, even when the agency had no questions.

In instances where the FDA has recommended reproductive toxicity tests on additives, such studies do not appear to have been done in the vast majority of cases.[32] And when health and safety studies indicate possible problems, food companies are not obligated to notify the agency except in very limited circumstances.[33]

30. Maffini, Looking Back, 449.
31. GAO, FDA Should Strengthen, page 20.
32. Neltner, Data Gaps (in press), 29.
33. Neltner, Navigating, 358.

The FDA also lacks clear authority to order companies to test the safety of chemicals they add to food. So even if the agency wanted to require additional testing, it is not clear that companies would be required to comply with such requests.[34] Further, companies are not required to regularly report the amount of a chemical added to food, making it difficult for the FDA to assess exposure or identify troubling use trends.[35]

The FDA's lack of authority to get the information it needs stands in stark contrast to the Environmental Protection Agency. Although its authority is still limited, the EPA has access to more information on pesticides and on chemicals that go into consumer products not regulated by the FDA than that agency has on chemicals used in food. Congress gave the EPA the ability to get information on pesticides and to make decisions on individual chemicals in a streamlined fashion that reduces the administrative burden without limiting transparency. This differential treatment of chemicals, which often are regulated simultaneously by both agencies, makes little sense.

As a result of these limits, thousands of chemical additives approved before 1980 have not been reassessed for safety.[36] With a lack of resources, no mandate from Congress, and an unusually difficult rule-making process, the FDA takes a passive approach to reviewing the safety of existing chemicals. Its failure to set science-based priorities for reassessment wastes resources, leads to litigation, undermines public confidence, and may result in firms selling unsafe food. Our analysis confirms the GAO's conclusion that "FDA is not systematically ensuring the continued safety of current GRAS substances" but also finds that it applies to most additives other than GRAS substances as well.[37]

Outdated Science

Much of the science that the FDA uses to review the safety of chemicals added to food has not been significantly updated for decades.[38] Using the 1982 report by its Select Committee on GRAS Substances as a baseline, we identified areas of concern where the issues raised more than 30 years ago remain unresolved and relevant today. They include the following:

- *Behavioral effects.* The FDA has not aggressively pursued the development of test methodologies for the impact of additives on behavior. It has not incorporated into its guidance methods that the EPA and other developed-country

34. Neltner, Data Gaps (in press), 35, 38.
35. Neltner, Navigating, 359; Neltner, Data Gaps (in press), 35.
36. GAO, FDA Should Strengthen, 20. If it has not done a systematic reassessment, then additives approved before 1980 were not reassessed. Many more than 4000 were approved by FDA before 1980; Alan M. Rulis, David G. Hattan, Victor H. Morgenroth 3rd, "FDA's Priority-Based Assessment of Food Additives. I. Preliminary Results," Regul Toxicol Pharmacol. 4 (1984):40.
37. GAO, FDA Should Strengthen, 20.
38. Maffini, Looking Back, 447; Maffini, Enhancing, 322.

members of the Organization for Economic Cooperation and Development adopted years ago.

- *Endocrine systems*. The FDA has not taken a leadership role in the development and validation of new technologies to identify and evaluate additives for potential endocrine disruption to hormones. Unlike the EPA, it has not adopted or made use of validated screening tests and predictive models.
- *Subpopulations*. The FDA has not systematically considered exposures of additives to sensitive populations except for infants. For hypersensitivity, it has not developed guidelines to screen or test for potential dangers or offered an effective system for consumers to report health problems.
- *Thresholds of alleged toxicological insignificance*. The FDA has adopted inadequate thresholds of exposure in rules and guidance below which industry is not expected to develop toxicity data when evaluating the safety of a chemical.
- *Absorption, distribution, metabolism, and excretion data*. The FDA's guidance allows industry to make safety decisions without the detailed data necessary to understand how the human body handles and eliminates chemicals that may be in food.
- *Reassessment and consistency across substances*. The FDA has not developed a system to prioritize its review of previous safety decisions. Instead, it relies on a case-by-case approach. In addition, it does not appear to closely coordinate its hazard or exposure assessment with the EPA when a chemical is regulated by both agencies.
- *Weight of evidence*. The FDA maintains that it closely scrutinizes all available studies. But its analysis is often based on professional judgment without using the available methods to compare various studies in a more rigorous, transparent, and reproducible manner.[39]

In addition, the program Congress imposed on the FDA in 1997 to review voluntary notifications for food-contact substances lacks transparency. Until the agency takes final action on a notice, the public is unaware of the decision, and the notices are not publicly available. This process limits participation by academics, competitors, public interest organizations, and the public in additive safety reviews.[40] As a result, the agency's decisions generally do not benefit from outside expertise.[41]

Missing Safety Deadlines

As noted, under a provision in the FDA Food Safety Modernization Act of 2011, industry will have to evaluate its processes to ensure that no unapproved food additives are used in its products. Yet the FDA has fallen behind in finalizing

39. Maffini, Looking Back, 446-447.
40. Neltner, Navigating, 359, 361; Maffini, Looking Back, 449.
41. Neltner, Navigating, 361.

those rules, which were supposed to take effect by July 2012, but were not even issued in draft form until January 2013. Until the law's regulations are in place, the FDA lacks an effective system to ensure compliance with food additive regulations, and food firms do not feel obligated to have internal management standards in place to prevent violations.[42] In the meantime, the FDA relies on tips and complaints from competitors, voluntary reports from manufacturers, or the infrequent inspections it conducts to identify compliance problems with its regulations for additives to food.[43]

PEW'S RECOMMENDATIONS

The systemic problems plaguing the food additive regulatory program prevent the FDA from ensuring the safety of all chemicals added to our food as Congress has intended, but there is no need to start over. Rather, it is better to adopt administrative and legislative solutions so that the Food and Drug Administration can more effectively ensure that new and existing uses of chemical additives in food are safe.

We recommend that the FDA take immediate action on its own to narrow the "generally recognized as safe" exemption to what Congress originally intended so that it is no longer a loophole and that it modernize its food additive science. We believe Congress should provide the FDA with the funding and authority it needs and ensure that the agency takes these actions. Legislative oversight and direction are essential to build and maintain stakeholder support.

For additives already on the market, the situation is different. The FDA lacks the clear authority to get the information it needs to identify problems, set priorities, and, when necessary, efficiently restrict the use of these additives to ensure safety. With about half of the additives already regulated by the Environmental Protection Agency under the Toxic Substances Control Act, and with many additives also regulated as pesticides, Congress needs to look at what is working at both agencies and devise an integrated chemical safety program that allows each agency to fulfill its essential responsibilities in a coordinated manner to minimize duplication of effort, record keeping, and the regulatory reporting burden on industry.

To accomplish these objectives, we make the following recommendations:

- Close the GRAS loophole.
- Modernize the FDA's food additive science.
- Ensure that existing chemical additives are safe.
- Establish a fee-based funding program to pay for the review process.

Each recommendation is examined in more detail below.

---

42. Neltner, Navigating, 367. Despite the statutory deadline, FDA has not finalized the prevention rules required by FSMA.
43. Neltner, Navigating, 357.

Close the GRAS Loophole

Congress should amend the Food Additives Amendment of 1958 so that the FDA approves the use of all new chemicals added to food and reviews significant changes to the use of previously approved additives by implementing a more streamlined and efficient decision-making process.

The food industry (and, to some extent, the agency itself) relies on the GRAS exemption because it believes that the food additive petition process that Congress adopted in 1958 is too burdensome and time-consuming, requiring that the FDA use extensive formal rule-making procedures. It prefers the GRAS notification program because it is informal and, as currently constructed, is voluntary. Manufacturers have the flexibility to seek FDA review when they want the legitimacy provided by that review and have the option to keep their innovations secret from competitors, even if the FDA and the public are excluded.

Yet as discussed above in our findings section and documented in our paper in the Journal of the American Medical Association-Internal Medicine, the GRAS program has serious problems, especially with regard to conflicts of interest, that must be addressed in order for the public to have confidence in the safety of foods.[44] We recommend that Congress require the FDA to review and, if appropriate, approve the first food use of a new chemical. The GRAS notification and the food-contact substance notification program should be limited to changes in existing uses or to additional uses only after the FDA has approved the chemical's use in food.

We recognize that concerns have been raised about the burden of the current food additive petition process. This is why we recommend that Congress establish a modernized, streamlined FDA approval process that the agency can effectively and efficiently administer; allows meaningful public input; meets industry's legitimate needs for timeliness and predictability; and, most importantly, restores public confidence that chemical additives are safe. Congress should consider as potential models existing programs that review and approve medical devices, drugs, and pesticides used on food.

To ensure that the notification programs are transparent and credible, Congress should allow the FDA to revise the food-contact substance notification program so it is more transparent. The notices, with confidential business information removed, should be publicly available on its website before the agency takes final action. Its decision letter to the company should also be posted on its website. This approach would be similar to what is done for GRAS notifications. In both programs, the public and stakeholders should have an opportunity to comment in an informal process before the agency takes final action.

---

44. Thomas G. Neltner, Heather M. Alger, James T. O'Reilly, Sheldon Krimsky, Lisa A. Bero, Maricel V. Maffini, "Conflicts of Interest in Approvals of Additives to Food Determined to be Generally Recognized as Safe: Out of Balance," Journal of American Medical Association–Internal Medicine, E2. Accessed August 13, 2013 (in press). doi: 10.1001/jamainternmed.2013.10559.

Until Congress changes the law, the FDA should revise how it implements the GRAS program to minimize conflicts of interest and to ensure that an additive's use is truly generally recognized as safe by the scientific community.

General recognition of additive safety requires consensus in the scientific community. There can be no such consensus if the chemical's use is unknown to the scientific community and to the FDA. The experts charged with assessing whether a scientific consensus exists should not have a relationship with the company that makes and sells the chemical additive. In addition, the experts must fairly represent the diversity of the scientific community.

Congress made clear that, in cases where there is no general recognition of safety, the FDA should make the decision through the food additive petition process, which requires public notice and comment followed by an FDA rule. As an interim step, the FDA should use its existing authority to establish clear guidance that an evaluation by an expert with a conflict of interest will not be effective

- Until the agency has reviewed it and agreed upon its safety.
- Unless the expert making the decision would be eligible to serve on an agency advisory committee considering the issue.

Modernize the FDA's Food Additive Science

Congress should ensure that the FDA uses the latest scientific methods to assess additive safety.

Consistency is important in science, but the FDA's approach to safety assessment is significantly different from those used by the EPA and other agencies. The FDA has the authority to upgrade its regulatory science and is committed to doing that through its Advancing Regulatory Science Initiative, but that initiative primarily focuses on areas other than chemical additives to food. Some of the differences in the way the EPA and the FDA assess additive safety stem from the FDA's being subject to an outdated law while the laws for other chemicals regulated by the EPA are more recent.

To overcome these safety assessment problems, Congress should provide the FDA with directions to modernize its program so that it evaluates a wider array of important health effects and improves its ability to ensure that public health is protected. It should consider the standards used in the Food Quality Protection Act of 1996, especially the margins of safety needed for vulnerable populations such as children and pregnant woman and the evaluation of chemicals for potential endocrine disruption.

The FDA should modernize the regulatory science it uses to evaluate the safety of chemical additives.

The FDA should upgrade the science used to evaluate the safety of additives through its ongoing Advancing Regulatory Science Initiative. As a critical first step, the agency should seek advice from an independent scientific advisory body to guide its modernization efforts. It should also continue and enhance its evaluation of the Tox21 program, which shows significant promise in setting priorities. The evaluation should consider two aspects:

- *Upgrade the science*: The FDA should define what constitutes harm and test for potential endocrine disruption, behavior effects, and developmental neurotoxicity at all life stages, including, when appropriate, additional safety factors for children and pregnant women.
- *Improve the process*: The FDA should more clearly separate the evaluation of science from management decisions, minimize conflicts of interest and bias from industry evaluations, provide a clear process to assemble and evaluate the evidence, and harmonize its analysis to be consistent with other agencies, especially the EPA.

The FDA should adopt and implement a science-based program to systematically review existing chemicals.

The FDA's current approach of reacting to citizen petitions, industry notifications, and media reports regarding additives already on the market is ineffective. Citizen petitions languish, and the agency stretches its limited resources, shifting from one additive to another without necessarily resolving the underlying scientific challenges.

The agency needs to rejuvenate its system to set priorities using modern scientific tools. We recognize that its Chemical Evaluation and Risk Estimation System is designed to accomplish this goal, but the details are unclear. Specifically, it is not clear that the system incorporates the information developed from Tox21 or that it will be rigorously validated through a transparent process that engages stakeholders. Both are essential.

The agency's top priorities should be widely used additives, those for which it lacks data, and those that are the source of public health concerns. Additives that do not merit immediate review include those recently reviewed and accepted by the European Union or other international organizations, or those that the FDA moves to a lower priority based on the available evidence after some form of public input. Based on these priorities, the agency should establish a schedule to reassess chemicals in light of its resources. Additives that are designated as low-priority could be moved up based on new science or changing uses.

Although the FDA may not immediately have the resources or tools to fully implement the plan, Congress is unlikely to give it what it needs without a plan in place. Industry and the public may withhold support if they lack confidence in its likelihood of success.

Ensure That Existing Chemical Additives Are Safe

Congress should update the law to give the agency the ability to obtain the information it needs to set priorities and reassess the safety of existing additives.

To effectively manage 10,000 additives, the FDA needs to efficiently estimate consumer exposure, be alerted by industry to health and safety studies, and require testing. With about half the additives already regulated by the EPA, we recommend that Congress strengthen and amend the Toxic Substances Control Act so food manufacturing companies routinely report to the EPA via existing programs the information about the use of those additives also regulated by that law and notify the FDA through the EPA when it believes there is a substantial risk posed by an additive that was previously unknown.

In addition, Congress should provide clear authority for the FDA to issue orders requiring the food industry to conduct testing and submit safety and use data to the agency, as the EPA is authorized to do for pesticides used on food.

The law should be updated to require the FDA to conduct a retrospective assessment of previously cleared chemicals through a transparent public process that sets priorities based upon available information.

Because 10,000 chemicals are already allowed in food, we acknowledge that it is not practical to conduct a thorough safety evaluation of all of them under current constraints. If the FDA had the authority to get the information it needs and coupled it with modern scientific tools to set priorities, then reviews could be done more quickly and efficiently—especially if food manufacturers cooperated in the analysis.

The task is daunting and would take time, but the increasing complexity of our food supply chain makes it necessary so that consumers can have confidence in it. We recommend that Congress and the agency set a specific timetable for designing and completing a review cycle, as was done for pesticide food tolerances under the Food Quality Protection Act of 1996.

Until Congress updates the law, FDA needs to request missing information from other agencies and industry.

Until Congress gives FDA the additional data collection authority and streamlined decision-making process it needs, we recommend that the agency obtain data from the EPA and from the European Commission, which has undertaken a similar effort for chemicals added to food and other consumer products. The agency should also request that industry provide it with all relevant health and safety studies and exposure information.

Establish a Fee-Based Funding Program

Congress needs to establish a fee-based program similar to that used for the pharmaceutical and pesticide industries to pay for the FDA's review and implementation of the food additives program. Although fees are not popular with manufacturers, the FDA otherwise will not be able to make the investment it needs to ensure that food additives are safe and to restore public confidence in the safety of these additives. The agency has sufficient experience with fees to make independent science-based decisions in a timely and effective manner.

Chapter 13

Protecting Consumers through New Regulations, Inspection, and Verification under the U.S. Food Safety Modernization Act

Chris Waldrop

Food Policy Institute, Consumer Federation of America, Washington, District of Columbia, USA

Chapter Outline

INTRODUCTION

Foodborne illness is a serious public health problem in the United States. Each year 48 million Americans are sickened by foodborne illness and 3000 persons die.[1] Children, the elderly, and individuals with compromised immune systems are at greatest risk of illness. The societal costs of foodborne illness have been estimated at $77 billion per year in medical costs, lost wages, and pain and suffering.[2] In addition, foodborne outbreaks can have serious economic costs to food producers and processors.

Illness and death from foodborne disease is largely preventable. Yet according to annual data on foodborne illness rates collected by the Centers for Disease Control and Prevention, there has been limited progress in reducing foodborne

1. Scallan, E., Hoekstra, R. M., Angulo, F. J., Tauxe, R. V., Widdowson, M., Roy, S. L., Jones, J. L., Griffin, P. M., 2011. Foodborne illness acquired in the United States—major pathogens. Emerging Infectious Diseases 17(1), 7-15.

2. Scharff, R. L., 2012. Economic burden from health losses due to foodborne illness in the United States. Journal of Food Protection 75(1), 123-31.

Food and Drug Regulation in an Era of Globalized Markets. http://dx.doi.org/10.1016/B978-0-12-802311-2.00013-8

illness over the past decade.[3] Changes in the food supply have also occurred in recent decades, such as an increased reliance on imported foods, changing consumer eating habits, emerging foodborne pathogens, and new pathogen-food combinations. Yet the laws governing food safety have often been slow to change and adapt. Meanwhile, several high-profile nationwide foodborne illness outbreaks linked to foods such as spinach, peppers, peanut butter, and eggs demonstrated significant gaps in the U.S. Food and Drug Administration (FDA)'s regulatory authorities.

The Food Safety Modernization Act (FSMA), signed into law in January 2011, was intended to address some of these food safety problems, reduce the toll of foodborne illness, and fill some of the regulatory gaps that have arisen in recent years.[4] The FSMA provides the FDA, which has oversight over 80% of the food supply, with new responsibilities and authorities to better assure the safety of the foods the agency regulates. Importantly, the FSMA shifts the FDA's approach from reaction—addressing problems only after contaminated food is on the market and consumers have become sick—to an approach focused primarily on prevention.

Under the FSMA, the FDA has the authority to require food processing plants to adopt systems of preventive controls that require plants to identify hazards (such as foodborne pathogens) in their food processing systems and take steps to prevent those hazards from occurring. The FSMA provides the FDA with new authority to require farms to take steps designed to safely produce fresh fruits and vegetables. The FDA has new authorities under the FSMA to better assure the safety of food imports, including holding importers accountable for importing safe food into the United States and establishing a third-party certification program for certain foods. The FSMA provides the FDA with a new inspection mandate—high-risk plants must now be inspected once every 3 years while lower-risk plants are to be inspected once every 5 years.[5] Other provisions of the new law include a requirement to develop an enhanced traceability system for high-risk foods, authority to develop new pathogen reduction performance standards, and authority to require companies to recall contaminated food products.

Consumer groups were actively involved in passage of the FSMA through Congress and are broadly supportive of the FDA's new preventive approach to food safety. As the FDA develops its implementing regulations, consumer

3. Crim, S. M., Iwamoto, M., Huang, J. Y., Griffin, P. M., Gilliss, D., Cronquist, A. B., Cartter, M., Tobin-D'Angelo, M., Blythe, D., Smith, K., Lathrop, S., Zansky, S., Cieslak, P. R., Dunn, J., Holt, K. G., Lance, S., Tauxe, R., Henao, O. L., 2014. Incidence and Trends of Infection with Pathogens Transmitted Commonly Through Food—Foodborne Diseases Active Surveillance Network, 10 U.S. Sites, 2006-2013. Morbidity and Mortality Weekly Report 63(15), 328-332.

4. FSMA does not address food safety issues related to meat and poultry products, as those products are regulated by the Food Safety and Inspection Service at the U.S. Department of Agriculture.

5. Previous inspection frequencies were not mandated by law; consequently FDA inspections occurred once every five to ten years.

groups have identified some key issues that must be resolved as the FDA finalizes its proposals. Several of those key issues are highlighted below.

PREVENTIVE CONTROLS FOR HUMAN FOOD

Under the agency's preventive controls proposal, food facilities will be required to develop and implement preventive controls in their facilities to prevent or mitigate the risk of contaminated food.[6] In order to ensure that facilities are developing adequate food safety plans, the FDA should obtain sufficient information from facilities at the outset to understand whether risks are being adequately identified and mitigated. While the FDA maintains that such information from hundreds of thousands of facilities across the world would be too voluminous to be useful, certain specific information could help inform allocation and prioritization of inspection resources and provide the agency with some sense of whether the food facilities are complying with the regulations.

The FDA should also require facilities to develop robust testing programs to help verify that facilities' food safety programs are producing safe food. Under the FDA's proposal, only facilities that produce certain ready-to-eat products would be required to conduct specific testing.[7] All other testing programs would be optional. Consumer groups believe that testing serves a valuable verification function and that facilities should have to meet robust testing requirements. This is especially important because the FDA's inspection resources are limited and the agency will not have a regular oversight presence in food facilities, instead inspecting facilities on an infrequent (once every 3 or 5 years) basis.

PRODUCE SAFETY

For first time ever, the FDA will require growers to follow basic safety standards in the production of fresh fruits and vegetables, including new standards for the safe use of water and manure, standards regarding animal intrusion and land use, and requirements to facilitate adequate worker hygiene.[8] These areas are important areas to address because each can potentially serve as a vector for the introduction or amplification of pathogen contamination on fresh produce if not appropriately managed.

6. Food and Drug Administration, 2013. Current Good Manufacturing Practice and Hazard Analysis and Risk-Based Preventive Controls for Human Food. Federal Register Vol. 78, No. 11, 3646-3824.
7. Food and Drug Administration, 2014. Current Good Manufacturing Practice and Hazard Analysis and Risk-Based Preventive Controls for Human Food. Federal Register Vol. 79, No. 188, 58524-58572, 58574-58599.
8. Food and Drug Administration, 2013. Standards for the Growing, Harvesting, Packing, and Holding of Produce for Human Consumption. Federal Register Vol. 78, No. 11, 3504-3646.

The FDA proposed to exempt certain produce that is "rarely consumed raw," reasoning that certain products like beets or sweet potatoes are typically cooked, which would reduce the risk of contamination.[9] However, consumer trends can change quickly and previous product consumption patterns may not predict future trends (i.e., certain foods that were not eaten raw 5 years ago may be eaten raw today). Cross-contamination in kitchens could also be exacerbated if certain products are not required to meet safety standards. Further, this proposal runs contrary to the basic principle and general consumer expectation that all food producers should produce safe food, regardless of the type of food. Consumer groups believe that no one should be exempt from food safety and that all types of produce should be covered under the regulations.

The most controversial issues in the FDA's new regulations deal with the use of water and manure. As part of the agency's regulations on water use, the FDA established a quantitative measure for water safety based on the Environmental Protection Agency's recreational water standard.[10] Though industry groups opposed the standard,[11] consumer groups supported the establishment of a numerical standard against which farms could measure their water quality. More problematic for the FDA was the establishment of a standard for manure. The FDA cited the need for additional research regarding raw manure use and minimum application intervals and suggested further study on the issue before establishing a specific standard in the regulations.[12] Consumer groups raised concerns about that approach and urged the agency to limit the length of time it would study the issue before developing specific requirements. In particular, consumer groups, in conjunction with sustainable agriculture advocates, urged the FDA to adopt an interim standard consistent with manure application intervals in the National Organic Program.

FOREIGN SUPPLIER VERIFICATION PROGRAM

Imported foods make up approximately 15% of the average American's diet, and in some categories of FDA-regulated foods, imported products comprise the majority.[13] About 80% of seafood consumed in the United States is imported; and about 50% of fresh fruits and 20% of fresh vegetables come from foreign countries.[14] Food and feed imports originate from more than 250,000 foreign establishments in 200 countries each year.[15] Food imports have grown

9. Food and Drug Administration, 2013. Standards for the Growing, Harvesting, Packing, and Holding of Produce for Human Consumption. Federal Register Vol. 78, No. 11, 3504-3646.
10. Food and Drug Administration, 2014. Standards for the Growing, Harvesting, Packing, and Holding of Produce for Human Consumption. Federal Register Vol. 79, No. 188, 58434-58473.
11. United Fresh, 2013. United Fresh comments to FDA on Standards for the Growing, Harvesting, Packing, and Holding of Produce for Human Consumption.
12. Food and Drug Administration, 2014. Standards for the Growing, Harvesting, Packing, and Holding of Produce for Human Consumption. Federal Register Vol. 79, No. 188, 58434-58473.
13. Food and Drug Administration, 2011. Pathway to Global Product Safety and Quality.
14. Food and Drug Administration, 2011. Pathway to Global Product Safety and Quality.
15. Food and Drug Administration, 2012. FDA FY 2013 Budget—Transforming Food Safety.

by an average of nearly 10% annually from 2002 to 2009.[16] However, the FDA is only able to inspect a small fraction—2%—of the foods that are imported into the United States.[17]

Under the FSMA, the FDA has new authority to hold food importers accountable for ensuring that their suppliers in foreign countries are providing them with safe food. The FDA's primary mechanism is a requirement that importers develop and maintain a Foreign Supplier Verification Program.[18] The verification component of the program is essential in order to provide importers with assurances that their supplier is meeting its obligations to provide safe food. Consumers groups believe that annual on-site verification of an importer's suppliers is critical as an on-site review can provide much greater assurance than a paperwork check. Under the new regulations, the FDA will require annual on-site reviews for certain foods that present a risk of serious adverse health consequences or death, but will provide importers with the option to forego annual audits if they can demonstrate that another verification measure is equally effective.[19] Consumers groups question whether importers will actually adopt the more robust measure of on-site audits if they are permitted to use other verification measures and whether those other measures can truly provide a similar level of assurance.

THIRD-PARTY CERTIFICATION PROGRAM

Another component of the FSMA's approach to imports is the development of a third-party certification program, under which the FDA would rely on third-party certifiers to verify that certain foods from foreign companies are being produced safely. While third-party certification is common in the food industry, it has been developed outside the regulatory context. As envisioned under the FSMA, the FDA will oversee third-party certification bodies that conduct specific activities as spelled out by the FDA.[20] This is a new program for the FDA and the agency has tried to align its approach with current industry schemes already in existence; though the law contains a few specific provisions that differ from industry practices. These differences are intended to better protect the public, such as a requirement that all audits are unannounced and a provision specifying that auditors report to the FDA any findings of public health significance during their audits.

16. Food and Drug Administration, 2013. Annual Report to Congress on Food Facilities, Food Imports and FDA Foreign Offices Provisions of the FDA Food Safety Modernization Act.
17. Food and Drug Administration, 2013. Annual Report to Congress on Food Facilities, Food Imports and FDA Foreign Offices Provisions of the FDA Food Safety Modernization Act.
18. Food and Drug Administration, 2013. Foreign Supplier Verification Programs for Importers of Food for Humans and Animals. Federal Register Vol. 78, No. 145, 45730-45779.
19. Food and Drug Administration, 2014. Foreign Supplier Verification Programs for Importers of Food for Humans and Animals. Federal Register Vol. 79, No. 188, 58574-58599.
20. Food and Drug Administration, 2013. Accreditation of Third-Party Auditors/Certification Bodies to Conduct Food Safety Audits and to Issue Certifications. Federal Register Vol. 78, No. 145, 45782-45839.

Consumer confidence in the third-party certification system overall has been damaged by several high-profile foodborne illness outbreaks in which domestic food producers were given high marks by third-party auditors only to be linked to huge nationwide outbreaks weeks later.[21] Consequently, transparency will be essential to the credibility of the FDA's third-party certification program. Providing the public with robust information about the entities involved in the program as well as posting regulatory audit reports and self-assessment reports will enhance transparency efforts. Further, adequate oversight of the program is essential—the FDA must ensure adequate funding and capacity to effectively oversee the program, including regular on-site audits of the entities engaged in the system.

ENFORCEMENT AND FUNDING OF THE FSMA

Finalizing the regulations is only the first step in realizing the full, intended benefits of the FSMA. The FDA then must implement the rules and ensure adequate oversight of food processors, food producers, importers, and third-party certifiers. Food inspections play an essential role in oversight so that the FDA can determine whether food processors, producers, and importers are complying with the new regulations. The agency will work closely with other entities, like state regulatory agencies and foreign governments, to ensure compliance with its regulations. Effective and efficient information systems will be critical so that the FDA can make informed decisions. Investments in technical capacity and training of agency personnel will be required. The FDA will have to develop guidance documents and technical assistance for the food industry, particularly small entities.

All of these activities require sufficient funding to be successful. The FDA needs adequate resources in order to ensure full implementation of the new law. Congress has recognized the important new role the FDA must play under the FSMA and has provided the agency with some additional money specifically dedicated for FSMA efforts. Particularly in a tough budget climate, this support for the FDA is highly welcome. However, to see the full benefits of the law as Congress intended, substantially greater appropriations will have to be provided. According to agency calculations, the FDA requires $400 to $450 million over 5 years to effectively implement the FSMA.[22]

The success or failure of the FSMA can be judged in many ways, though the most important measure is whether or not activities under the FSMA reduce foodborne illness and death. Direct correlation is complicated by the complexity of the food system, lack of specific data, and multiple activities and interventions being implemented at once. Yet improved food safety requirements, adequate oversight, and sufficient funding are clearly critical components of the FDA's new preventive approach to food safety.

21. For example: Peanut Corporation of America, Jensen Farms, Wright County Eggs.

22. Food and Drug Administration, 2013. Report to Congress on Building Domestic Capacity to Implement the FDA Food Safety Modernization Act.

Chapter 14

The Evolving Role of the Codex Alimentarius Commission in Securing Consumer Protection in Globalized Food Markets

Aliza Y. Glasner
O'Neill Institute for National and Global Health Law, Washington, District of Columbia, USA

Chapter Outline

When formulating national policies and plans with regard to food, Governments should take into account the need of all consumers for food security and should support and, as far as possible, adopt standards from the Food and Agriculture Organization's... and the World Health Organization's Codex Alimentarius...
—1985, United Nations General Assembly, Guidelines for Consumer Protection

Since its founding in 1963, the Codex Alimentarius Commission (Codex) has served as a global standard-setting organization, presiding over the safety and quality of food across the world. Founded by the World Health Organization (WHO) and the Food and Agriculture Organization (FAO), Codex established internationally recognized food standards, guidelines, and codes of practice.[1] The desired outcome of Codex work is a world where consumers are able to

1. *About Codex*, Codex Alimentarius, http://www.codexalimentarius.org/about-codex/en (last updated Aug. 10, 2014).

Food and Drug Regulation in an Era of Globalized Markets. http://dx.doi.org/10.1016/B978-0-12-802311-2.00014-X

rely on the safety and quality of the food they buy and importers can trust that deliveries meet at least minimal, uniform international standards.[2]

Codex promulgates nonbinding, voluntary standards intended to reflect international consensus as to components, products, processes, and labeling.[3] As the "global reference point" for food quality standards, consumers, food producers and processors, national food control agencies, and the international food trade all rely upon Codex work.[4] Codex standards cover everything from product identification to packaging specifications and play an important role in foods moving in international commerce. For example, canned tuna fish, a popular food item the world over, may make five stops between harvesting and a retailer's shelf (Figure 14.1).[5] According to Codex, there are 14 species of fish definable colloquially as "tuna" around the globe. The six-page standard discusses

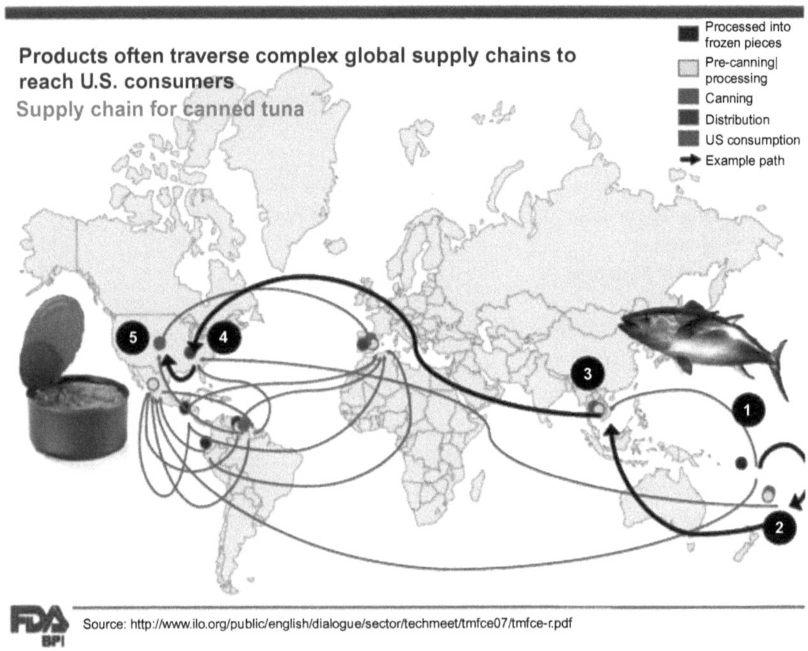

FIGURE 14.1 Illustrative supply chain for canned tuna.

2. *Id.*
3. *Id.*
4. Food and Agriculture Organization of the United Nations & World Health Organization, *Understanding the Codex Alimentarius* 14 (3d ed. 2006) [hereinafter Understanding the Codex], ftp://ftp.fao.org/codex/Publications/understanding/Understanding_EN.pdf.
5. *See infra* Exhibit 7.

everything from the hermetic seal on the can to the level of allowable contaminants in each can.[6,7]

Codex has three core functions: to provide scientific advice, establish food standards, and create capacity building programs.[8] Scientific advice is solicited and shaped by both permanent and *ad hoc* committees charged with analyzing discrete products and processes.[9] Based upon the work published by general subject, commodity, *ad hoc*, and regional committees, Codex develops standards. Then the FAO and the WHO support the standard-setting work undertaken by Codex committees and create capacity building programs on food safety and quality informed by Codex work.[10]

Codex is driven by two, arguably competing missions: to adopt standards that protect consumer health and that promote fair trade.[11] This latter mandate has ascended in importance since 1995, and, at least according to critics, has done so to the detriment of its public health charge. Increased globalization and the changing trade and investment climate, have exerted unprecedented and considerable pressure on Codex's process for scientific assessment and standard development. As the world's food supply becomes increasingly global, reliance on and confidence in harmonized standards is essential for efficient international trade. The rise of the World Trade Organization (WTO) in particular changed Codex's role on the global stage, ostensibly giving Codex standards stronger legal effect and at the same time further marginalizing states already disenfranchised from Codex for political and economic reasons.[12] Given that Codex's reach spans over 99% of food consumed around the world, its standards portend enormous influence on both health and trade.[13]

This chapter explores this tension by exploring first Codex's structure and then how that structure has become susceptible to greater outside governmental

---

6. Codex Alimentarius, *Codex Standard for Canned Tuna and Bonito*, CODEX STAN 70-1981 (1981) (*rev'd* 1994, *amended in* 2011 & 2013), *available at* http://www.codexalimentarius.org/download/standards/105/CXS_070e.pdf.

7. Office of Global Regulatory Operations & Policy, U. S. Food & Drug Admin., *Pathway to Global Product Safety and Quality* (*rev'd* 2011), *available at* http://www.fda.gov/downloads/AboutFDA/CentersOffices/OfficeofGlobalRegulatoryOperationsandPolicy/GlobalProductPathway/UCM262528.pdf.

8. Food and Agriculture Organization of the United Nations & World Health Organization, *Understanding the Codex Alimentarius* 14 (3d ed. 2006) [hereinafter *Understanding the Codex*], *available at* ftp://ftp.fao.org/codex/Publications/understanding/Understanding_EN.pdf.

9. *List of All Codex Committees*, Codex Alimentarius, http://www.codexalimentarius.org/committees-and-task-forces/en/?provide=committees (last accessed Jan. 20, 2015) [hereinafter *Codex Committees*].

10. *Understanding the Codex*, *supra* note 8, at 14.

11. "When formulating national policies and plans with regard to food, Governments should take into account the need of all consumers for food security and should support and, as far as possible, adopt standards from the … Codex Alimentarius or, in their absence, other generally accepted international food standards". G.A. Res. 39/248, U.N. Doc. A/RES/39/248 (Apr. 16, 1985), *available at* http://www.un.org/documents/ga/res/39/a39r248.htm.

12. Michael A. Livermore, *Authority and Legitimacy in Global Governance: Deliberation, Institutional Differentiation, and the Codex Alimentarius*, 81 N.Y.U. L. Rev. 766, 783 (2006).

13. *Understanding the Codex*, *supra* note 8, at 21.

and nongovernmental influence after the rise of the global trade regime overseen by the WTO. Specifically, this chapter focuses on the role of outside influences and potential negative implications for consumer protection and equity in ensuring a globally safe food supply.

THE INFRASTRUCTURE OF THE CODEX ALIMENTARIUS COMMISSION

A Brief History

The need for an organization like Codex was created by dueling concerns at the international level: the establishment of domestic standards for food safety and the resulting trade barriers presented as conflicting standards arose.[14] Beginning in the 1940s, increasing consumer knowledge about food generally and more specifically about food safety led governments to adopt measures to protect the public from potentially dangerous and low-quality food.[15] Food regulators, traders, and experts looked to the FAO and the WHO to assist with the increasing number and complexity of food regulations, which were creating trade barriers without effectively protecting consumers.[16]

Joint meetings between the FAO and the WHO on nutrition and food additives began in 1950, and shortly after the WHO recognized the increasing use of chemicals in food as a public health problem.[17] Concurrently, several countries proposed regional food codes in Latin America and Europe.[18] In 1960, the first FAO Regional Conference for Europe recommended a joint FAO/WHO program on food standards and supported the establishment of international minimum food standards.[19] The 1962 FAO/WHO Food Standards Conference requested the implementation of an international food standards program and created the Codex Alimentarius.[20]

ORGANIZATIONAL STRUCTURE

Membership in Codex is voluntary and open to members of both the WHO and the FAO.[21] Membership includes over 180 countries, and one organization—the European Union (EU). Only members may vote to approve standards and guidelines, sit on committees, or nominate candidates for executive positions.[22]

14. Ching-Fu Lin, *Global Food Safety: Exploring Key Elements for an International Regulatory Strategy*, 51 Va. J. Int'l L. J. 637, 671 (2011).
15. *Understanding the Codex*, *supra* note 8, at 14.
16. *Id.*
17. *Id.* at 13.
18. *Id.*
19. *Id.*
20. *Understanding the Codex*, *supra* note 8, at 13.
21. *Codex Members and Observers*, Codex Alimentarius, http://www.codexalimentarius.org/members-observers/en/ (last updated May 14, 2014) [hereinafter *Codex Members and Observers*].
22. *Id.*

A country, organization, intergovernmental organization (IGO), and nongovernmental organization (NGO) that does not want or does not qualify for membership status may participate as an observer party (observers).[23] Many international organizations and consumer, environmental, industry, and nongovernmental organizations participate in Codex as observers.[24] Observers must request an invitation from the Codex Secretariat to attend sessions or *ad hoc* meetings of either the commission or its subcommittees.[25] Though observers are not formally permitted to give input at sessions, they are often invited to inform the discussion and offer opinions.[26] Nongovernmental organizations also participate through their national delegations.

LEADERSHIP

Codex Executive Committee

The Codex Alimentarius Commission serves as the organization's decision-making body and is "responsible for compiling the standards, codes of practice, guidelines and recommendations that constitute the Codex Alimentarius."[27] The Codex Executive Committee is responsible for meeting between sessions of the commission, and acts on behalf of the commission as the lead committee.[28] Since 1963, the commission has met at least annually to approve new work and adopt Codex texts prepared by Codex committees and task forces. The executive committee plays a key role in determining whether standards published by Codex are created or revised.[29] After a project proposal for a new standard is received, the executive committee reviews the proposal and evaluates the proposal against the criteria and priorities determined by the committee as a whole.[30] The executive committee determines whether a standard will proceed into development, which involves extensive consultation with governments and other interested parties, followed by committee debate.[31]

Codex Secretariat

The secretary of Codex is responsible for coordinating and administering the activities of the commission.[32] The Directors-General of the FAO and the WHO are responsible for appointing the Secretary of the Codex Alimentarius.[33] The secretariat is based at FAO headquarters in Rome, and is responsible for administering commission and executive committee meetings.[34] The secretariat

23. *Id.*
24. *Id.*
25. *Id.*
26. *Understanding the Codex*, *supra* note 8, at 22.
27. *Id.* at 14.
28. *Id.*
29. *Id.* at 16.
30. *Id.* at 16-17.
31. *Understanding the Codex*, *supra* note 8, at 16.
32. *Id.* at 19.
33. *Id.*
34. *Id.*

is also responsible for coordinating with the member governments who host, finance, and maintain the commission's subsidiary committees.[35] Member governments who are responsible for subsidiary committees collaborate with the secretariat to determine the timing and agenda for meetings, provide information to attendees, and ensure any follow-up activities are completed.[36]

Codex Subsidiary Bodies

The commission is divided into three subsidiary bodies, which prepare and coordinate standards and guidelines, including the general subject committees, commodity committees, FAO/WHO coordinating committees, and an *ad hoc* intergovernmental task force.[37] Codex committees are responsible for preparing draft standards in consultation with the executive committee, governments, and interested parties.[38] Coordinating committees are the organizations through which regions or other discrete groups of countries may coordinate food standard development in the region.[39] The commission designates the host member country for each committee, who is also responsible for the cost of maintaining and administering the committee, along with designating a chairperson for the committee.[40]

Each Codex member country sends designated representatives to serve on Codex subcommittees. Lower-income countries often rely on Codex as a means of adopting regulations, standards, and policies. For countries without strong food regulatory authorities, or without the proper internal infrastructure to adequately regulate food, Codex provides aid in the regulatory process although it cannot guarantee minimal participation for all interested members.[41] The success of applying Codex standards domestically relies on finding a country representative who is technically proficient but also understands the cultural, environmental, and political factors affecting food regulation.

This designee must have knowledge of Codex rules and procedures, a well-functioning national Codex contact point, and effective negotiation and strategy building skills.[42] The designee is responsible for working internationally with Codex as well as facilitating domestic integration of Codex standards into national regulation and policy.[43] For example, the designee must have national expertise in food safety and its regulation. Including:

- Understanding of and ability to apply the risk analysis framework;
- Experience implementing food control activities within a well-managed food control system;

35. *Id.*
36. *Understanding the Codex*, *supra* note 8, at 19.
37. *Codex Committees*, *supra* note 9.
38. *Id.*
39. *Understanding the Codex*, *supra* note 8, at 16.
40. *Id.*
41. Food and Agriculture Organization of the United Nations, *Facilitating Effective Participation in CODEX*, *available at* http://www.fao.org/fileadmin/user_upload/agns/pdf/factsheets/Codex.pdf.
42. *Id.*
43. *Id.*

- Technical/scientific expertise in different food safety-related disciplines;
- Access to adequate and reliable national food safety data;
- Mechanisms that assure effective communication with the food industry and with consumers on food safety issues;
- Mechanisms that optimize the involvement of the academic and research communities in food safety.[44]

The utility of a given standard is only realized when an adopting country has the capacity to implement and comply with the standard. Accordingly, an adopting country needs sufficient, strong, and enforceable internal food legislation as well as infrastructure for inspection and laboratory functions, technical, and administrative support. For high-income countries, Codex guidelines are heavily relied upon when adopting policies and writing national regulation. In the United States, for example, eight federal agencies, as well as many subdivisions of these agencies, regularly rely on Codex standards.[45]

Consequently, Codex must also play a strong role in capacity building for members in need of assistance. Capacity building programs happen at the national, subregional, and regional levels and often include programming-like seminars, training of trainers' courses, workshops, preparation of tools, as well as field projects and experts' missions.[46]

To support its capacity building activities and to address growing disparities in access to Codex processes, in 2003 the FAO and the WHO launched the FAO/WHO Project and Fund for Enhanced Participation in Codex (Codex Trust Fund).[47] The concept was to provide support to developing and transition economy countries to enhance their level of effective involvement in Codex through participation in meetings and submission of scientific data.[48] The original goal of the fund was to raise US$40 million over a 12-year period and distribute based on need. At the end of 2013, the fund raised approximately half of the stated goal.[49] India and Malaysia provide examples of countries that have transitioned from beneficiaries to donors with Codex Trust Fund support,

---

44. *Id.*
45. *See* Food Safety and Inspection Service, U.S. Dept. of Agric., *U.S. Codex Strategic Plan 2012-2017* (2012), *available at* http://www.fsis.usda.gov/wps/wcm/connect/ab59575e-0ccc-4ff5-8ecb-300a0bbf5ffb/US_Codex_Strategic_Plan_2012-2017.pdf?MOD=AJPERES.
46. Food and Agriculture Organization of the United Nations, *FAO/AGNS Capacity Building Activities Related to Codex*, *available at* http://www.fao.org/fileadmin/templates/agns/pdf/Codex_activities_by_regions_05_07.pdf.
47. Ravi Afonso Pereria, *Why Would International Administrative Activity Be Any Less Legitimate?—A Study of the Codex Alimentarius Commission*, 9 Ger. L.J. 1693, 1709 (2008).
48. *Codex Trust Fund: Background*, World Health Organization, http://www.who.int/foodsafety/areas_work/food-standard/codextrustfund/en (last accessed Jan. 20, 2015).
49. "At the end of 2013 contributions totaling US$ 18,770,007 had been received from 15 Codex Member States and one Member Organization. Countries who have made contributions to the Codex Trust Fund are as follows: Australia, Canada, European Commission, Finland, France, Germany, India, Ireland, Japan, Malaysia, Netherlands, New Zealand, Norway, Sweden, Switzerland, United States of America." *Codex Trust Fund: Donor Contributions*, World Health Organization, http://www.who.int/foodsafety/areas_work/food-standard/codextrustfund/en/index1.html (last accessed Jan. 20, 2015).

although there is widespread consensus that developing countries need substantially more assistance, beyond even a fully funded Codex Trust Fund.[50]

Codex Session Procedures

The commission meets in regularly scheduled sessions every 2 years, with the location alternating between the FAO headquarters in Rome and the WHO headquarters in Geneva.[51] Each member country is represented by national delegations led by senior officials appointed by the country's government.[52] Each delegation often includes representatives from industry, consumer organizations, and academic institutions. Other entities are permitted to attend in an observer capacity, including countries that are not yet members, international governmental organizations, and nongovernmental organizations.[53] Observer organizations may express opinions, but may not vote on the final decision.[54]

CODEX IN PRACTICE: A BRIEF LOOK AT SOME GLOBAL CHALLENGES

Internal Design

By design, Codex does not conduct independent scientific review, but rather relies on studies submitted by its members to set standards by consensus. Committees consider large amounts of data, and use an extensive deliberation process to develop standards.[55] This process inherently invites problems for a number of reasons. Submitted studies are typically industry funded and come from developed countries.[56] The work of the committees is centered on the needs of wealthier nations, which generally further or at least dovetail with trade advantages and economic gain. Codex critics have frequently asserted the potential and probable bias toward standards that favor wealthy countries' interests.[57]

Lower-income countries generally do not have the resources to contribute scientific research and expertise to Codex committees.[58] A 2002 independent evaluation of Codex participation found 96% of low-income countries and 87%

---

50. *Id.*
51. *Understanding the Codex*, *supra* note 8, at 15.
52. *Id.*
53. *Id.*
54. *Id.*
55. Livermore, *supra* note 14, at 783.
56. Kuei-Jung Ni, *Does Science Speak Clearly and Fairly in Trade and Food Safety Disputes? The Search for an Optimal Response of WTO Adjudication to Problematic International Standard-Making*, 68 Food & Drug L.J. 97, 98 (2013).
57. Livermore, *supra* note 14, at 783.
58. *Id.* at 773-74.

of middle-income countries were dissatisfied with their level of participation in Codex, citing financial issues as the leading barrier.[59] Concentrating decision making in the hands of those who can afford to fund research means lower-income countries that cannot afford to commission studies might lose an opportunity to be heard.[60]

OUTSIDE INFLUENCES

Other International Institutions

For much of Codex's history, the organization has served primarily to promote harmonized and reliable food standards and to provide a forum for sharing scientific information and regulatory best practices.[61] The advent of the WTO fundamentally altered Codex's trajectory, giving legal weight and considerable power to Codex standards.[62]

In 1995, during the Uruguay Round of trade negotiations, signatories to the WTO simultaneously bound their national laws to the WTO's Sanitary and Phytosanitary Agreement (SPS agreement).[63] The SPS agreement regulates food, plant, and animal safety and health by relying on Codex standards. Accordingly, Codex standards are codified into the system of rules and incentives at the WTO.[64] "When establishing or maintaining sanitary or phytosanitary measures to achieve the appropriate level of sanitary or phytosanitary protection, Members shall ensure that such measures are not more trade-restrictive than required to achieve their appropriate level of sanitary or phytosanitary protection, taking into account technical and economic feasibility."[65]

According to the FAO, "Codex standards have become the benchmarks against which national food measures and regulations are evaluated within the legal parameters of the WTO Agreements."[66] Consequently, for any

59. W. Bruce Traill et al., Food and Agriculture Organization of the United Nations & World Health Organization, *Report of the Evaluation of the Codex Alimentarius and Other FAO and WHO Food Standards Work* (Nov. 15, 2002), *available at* http://www.fao.org/docrep/meeting/005/y7871e/y7871e00.htm.

60. *Id.*; *see also* Ni, *supra* note 56, at 98-99.

61. Livermore, *supra* note 14, at 771.

62. *Id.* at 774.

63. The WTO's SPS Agreement was established in 1995 to regulate food, plant, and animal safety and health regulations. The adjudicatory arm ("Panels") of the WTO resolves trade disputes regarding such issues, and can impose or permit trade-based punitive measures for violations of the SPS Agreement. *See Understanding the WTO Agreement on Sanitary and Phytosanitary Measures*, World Trade Organization, http://www.wto.org/english/tratop_e/sps_e/spsund_e.htm (last visited Apr 5, 2014) [hereinafter *Understanding the SPS Agreement*].

64. Livermore, *supra* note 14, at 771.

65. Agreement on the Application of Sanitary and Phytosanitary Measures, Article 5, ¶6, The Legal Texts: The Results of the Uruguay Round of Multilateral Trade Negotiations 59 (2000), 1867 U.N.T.S. 493 (1994) [hereinafter *SPS Agreement*], at Article 5, Paragraph 6.

66. *Understanding the Codex*, *supra* note 8.

country to issue national standards that are more stringent than the Codex standards adopted by the SPS agreement, a country must submit an application with adequate scientific justification to the WTO.[67] Though technically speaking Codex standards remain neither obligatory nor self-executing, membership in the WTO renders them effectively binding. Therefore, by virtue of the SPS agreement, Codex's role in determining the appropriate relationship between consumer protection and trade liberalization gained unparalleled significance.

On its face, bolstering Codex standards with the legal weight of the WTO has positive implications for ensuring its mission to support a safer food supply. Under Codex's own mandate, its standards are reviewed domestically, without any international independent review process. Under the SPS agreement, any state can be more easily assured that a trading partner is going to adopt a particular Codex standard, implying more reliability for food trade.[68] The threat of a trade dispute over a state's refusal to respect a Codex standard, at least in theory, helps to ensure compliance.[69] Most significantly, the SPS agreement provides a means for international harmonization of national sanitary and phytosanitary measures, which has the potential to promote trade liberalization of agriculture and food products.[70]

The reality, however, is considerably more nuanced. While the SPS agreement has been good for trade generally, its influence has also brought unprecedented controversy to Codex and thrust the organization into an uncharted legal and political arena.[71] Critics point out that the agreement inherently prioritizes trade above national health determinants.[72] Under the SPS agreement, a national health standard is generally prohibited if the WTO decides that it is not based on scientific principles and is maintained without sufficient scientific evidence.[73] As a result, members have greater incentives to follow WTO standards, and feel more coercive pressure to comply.[74] The cost of litigation is extremely high and the financial penalties for noncompliance with WTO standards are prohibitive.[75] Adopting international standards is a rational and obvious way for countries to avoid litigation and penalties.[76]

---

67. *See SPS Agreement*, *supra* note 65, at 3.3, ¶5; Livermore, *supra* note 14, at 768-69, n.9.
68. *See SPS Agreement*, *supra* note 65, at Article 3.1.
69. Livermore, *supra* note 14, at 772.
70. *See* Ni *supra* note 56; *see also SPS Agreement*, *supra* note 65, at Preamble Recitals 5 & 6.
71. Livermore, *supra* note 14, at 772.
72. Bruce Silverglade, Center for Science in the Public Interest, The WTO Agreement on Sanitary and Phytosanitary Measures: Weakening Food Safety Regulations to Facilitate Trade (2000), *available at* https://cspinet.org/reports/codex/dutch_wto.html.
73. *SPS Agreement*, *supra* note 65, at Article 2, ¶ 2.
74. Livermore, *supra* note 14, at 772.
75. *Id.*
76. *Id.*

Industry

Industry's influence at the WTO and Codex is also pervasive. National governments are highly influenced by industry lobbying and this pressure reaches the WTO.[77] The SPS agreement provides a bargaining chip for one state (often under the influence of industry) to threaten suit against another for establishing more stringent or conflicting domestic standards that might limit trade in some capacity.[78]

Codex's process of determining standards by scientific consensus can be at odds with national tolerance for risk. Lack of agreement on risk tolerance has posed long-standing barriers to food trade.[79] This tension is well illustrated by a decades-old debate between the EU and the United States regarding the use of antibiotics to treat livestock in industrial agriculture. Using antibiotics to treat agricultural livestock for growth promotion and prophylactic health has well-established consequences for human health. Specifically, scientific evidence supports that "the use of antibiotics in humans and animals places individuals at increased risk of infection, higher treatment failures, and increased severity of illness."[80] At the same time, the international meat trade and the trade of the antibiotics themselves is enormously lucrative and therefore jeopardized by any sort of ban or restriction.

In 1999, the United States Trade Representative (USTR) challenged a EU ban on the use of human-use antibiotics in livestock feed claiming a violation of the SPS agreement. The complaint alleged that the EU had not completed a proper risk assessment.[81] The USTR's claims were met with skepticism because the U.S. government independently published scientific conclusions consistent with the EU's findings, specifically: (1) the U.S. Centers for Disease Control (CDC) had concluded that the EU ban is "scientifically justifiable" and "protects the public health"; (2) 2 years earlier, the WHO recommended a prohibition on human-use antibiotics to promote animal growth; and (3) the U.S. National Academy of Sciences had concluded in 1998 that "there is a link between the use of antibiotics in food animals, the development of bacterial resistance to these drugs, and human disease."[82]

---

77. Silverglade, *supra* note 72, at 785-86.

78. WTO, *The Impact of the TBT and SPS Agreements on Food Labeling and Safety Regulations*, Center for Science in the Public Interest, https://cspinet.org/reports/codex/wtospsbt.htm (last visited November 16, 2014) [hereinafter *Impact of TBT and SPS*].

79. David Vogel, The Politics of Precaution: Regulating Health, Safety and Environmental Risks in Europe and the United States 91 (2012).

80. *Id.* at 92 (*citing* The Pew Charitable Trust, *Save Antibiotics—The Campaign Fact Sheet, Campaign on Human Health and Industrial Farming, available at* http://www.saveantibiotics.org/ourwork.html (last accessed Jan. 22, 2015)).

81. *Id.*

82. Preservation of Antibiotics for Human Treatment Act, S. 2508, 107th Cong. § 2 (2002).

Still, the USTR pursued its action motivated by concern that the ban would threaten a lucrative livestock trade, as well as the trade of the antibiotics themselves.[83] Further, the USTR was concerned that a ban would cause a ripple effect on U.S. meat exports themselves, and stir consumer desires for labeling changes for meat from animals treated with antibiotics.[84] Twenty-five years later, the United States and the EU still maintain conflicting standards for antibiotic use in livestock.[85] The United States still allows antibiotics for the use of growth promotion in livestock, and the EU does not.[86]

In 2004, Japan explored imposing stronger regulation for human-use antibiotics in livestock. The United States pressured their decision making, again concerned that tighter standards would negatively impact U.S. livestock exports.[87] The use of antibiotics in livestock illustrates just one of many debates where scientific evidence is mired in politics and economics. Scientists broadly agree that using human-use antibiotics for livestock is dangerous for human health. Unfortunately, lucrative trade policy and public health considerations do not always align.

CODEX IN THE TWENTY-FIRST CENTURY

In the twenty-first century, Codex has an opportunity to morph its capacity to meet the ever-changing needs of an increasingly globalized and codependent world. Programs like the Codex Trust Fund provide concrete mechanisms to address the organization's functional challenges, such as meeting the needs of lower-income countries. Lower-income countries who both supply and rely on the international food trade must be ensured an equal voice and opportunity to contribute. While the SPS and the WTO play a tremendous role in encouraging harmonization of standards and adoption of safer food practices, their political

83. *U.S. Warns EU on Antibiotics Ban*, 3 BRIDGES Weekly Trade News Digest (Int'l Ctr. for Trade and Sustainable Dev., Geneva, Switz.), Sept. 13, 1999, at 6-7, *available at* http://www.ictsd.org/sites/default/files/review/bridgesweekly/bridgesweekly3-36.pdf.

84. *Id.*

85. *Id.*

86. In 2014, the Food and Drug Administration guidelines for industry to voluntarily phase out the non-therapeutic use of antibiotics in livestock. *See* U. S. Food & Drug Admin., *FDA's Strategy on Antimicrobial Resistance—Questions and Answers* (*rev'd* Mar. 28, 2014), http://www.fda.gov/AnimalVeterinary/GuidanceComplianceEnforcement/GuidanceforIndustry/ucm216939.htm (last accessed Jan. 22, 2015).

87. "The United States expressed its concerns to the Japanese government and sought assurances that Japan's review of these additives would be performed in a transparent, thorough, and science-based manner. The Japanese government provided such assurances, and the United States will continue to follow the issue closely to ensure that Japan decides this matter in a manner consistent with its WTO obligations." Office of the U.S. Trade Representative, Exec. Office of the President, 2004 National Trade Estimate Report on Foreign Trade Barriers 263 (2004), *available at* http://www.ustr.gov/archive/assets/Document_Library/Reports_Publications/2004/2004_National_Trade_Estimate/2004_NTE_Report/asset_upload_file231_4191.pdf.

might is a challenge to be understood and addressed. Scientific consensus and resulting regulatory standards must be achieved free from trade and political pressure.

Food production is no longer a national issue. A safe food supply requires increasing global regulatory harmonization and reliance on the efforts of foreign governments to comply. In order to carry on the mission of the founding tenets of the Codex Alimentarius, the organization must adapt and grow to meet the challenges now facing the world's food supply.

Chapter 15

The Emergence and Influence of Transnational Private Regulation of Food Safety*

Ching-Fu Lin
Institute of Law for Science and Technology, National Tsing Hua University, Hsinchu, Taiwan

Chapter Outline

*A previous version of this chapter was published by the Food and Drug Law Institute in Volume 69, Issue 2, of the Food and Drug Law Journal. The author thanks the Food and Drug Law Journal for the permission to adapt this Chapter from the publication.

Food and Drug Regulation in an Era of Globalized Markets. http://dx.doi.org/10.1016/B978-0-12-802311-2.00015-1

INTRODUCTION

Due to a regulatory void in the public regulation of food safety at local and global levels,[1] private actors—especially multinational food companies, supermarket chains, and nongovernmental organizations (NGOs)—are increasingly filling these gaps. These private entities employ private standards, certification protocols, third-party auditing, and transnational contracting practices to fill the gap.[2] The emergence of transnational private regulation (TPR)[3] in the food safety arena has been alongside the gradual decline of states' traditional command-and-control regulation. These state regulations are increasingly being supplemented or replaced by more flexible, market-oriented mechanisms.[4] Examples include Global Good Agricultural Practices (GlobalGAP), the British Retail Consortium (BRC) Global Standards, and Tesco's Nature's Choice, to name a few. Although TPR is, in theory, not mandatory for suppliers, many TPR schemes have a de facto mandatory status. The de facto mandatory status arises from a large number of buyers in global agri-food markets now requiring their suppliers to meet such private requirements, which are usually stricter and more prescriptive than their public counterparts. In particular, these private requirements are reinforced by the participation of players with enormous market power such as WalMart, Carrefour, and McDonald's. TPRs, as such, are able to "filter[] all the way back through processors, traders, and down to the level of agricultural production."[5] These TPRs are a testament to the increasing importance of private governance tools in the global regulatory space of food safety.

There has been a long history in regulatory activities by private actors in the area of food safety and quality at a *local* level.[6] The involvement of private actors in food-related regulatory activities can at least be dated back to the Middle Ages, where "trade guilds" in Europe were able to exercise a certain level of

---

1. *See generally* Ching-Fu Lin, *Global Food Safety: Exploring Key Elements for an International Regulatory Strategy*, 51(3) Virginia Journal of International Law 637 (2011).
2. *See e.g.* Jennifer Clapp & Doric Fuchs eds., Corporate Power in Global Agrifood Governance (2009). Denise Prévost, *Private Sector Food-Safety Standards and the SPS Agreement: Challenges and Possibilities*, 33 S. Afr. Y.B. Int'l L. 1 (2008); Linda Fulponi, *Private Voluntary Standards in the Food System: The Perspective of Major Food Retailers in OECD Countries*, 31(1) Food Pol'y 1 (2006).
3. For a broader analytical introduction, *see* Burkard Eberlein, Kenneth W. Abbott, Julia Black, Errol Meidinger and Stepan Wood, *Transnational Business Governance Interactions: Conceptualization and Framework for Analysis*, 8(1) Regulation & Governance 1 (2014).
4. *See generally* Bridget M. Hutter, Regulation and Risk: Occupational Health and Safety on the Railways 9-10 (2001); Neil Gunningham & Darren Sinclair, *Regulatory Pluralism: Designing Policy mixes for Environmental Protection*, 21(1) Law & Pol'y 49 (1999); Franco Furger, *Accountability and Systems of Self-Governance: The Case of the Maritime Industry*, 19(4) Law & Pol'y 445 (1997); Philip Eijlander, *Possibilities and Constraints in the Use of Self-Regulation and Co-Regulation in Legislative Policy: Experience in the Netherlands—Lessons to Be Learned for the EU?* 9(1) Electronic J. Comp. Law (2005); Philipp Pattberg, *Private Governance and the South: Lessons from Global Forest Politics*, 27(4) Third World Q. 579 (2006); John Braithwaite, *Enforced Self-Regulation: A New Strategy for Corporate Crime Control*, 80(7) Mich. L. Rev. 1466, 1466-83 (1982).
5. Clapp & Fuchs, *supra* note 2, at 14.
6. Fabrizio Cafaggi, *Private Regulation, Supply Chain and Contractual Networks: The Case of Food Safety*, EUI Working Paper RSCAS 2010/10 (2010), at 1.

control over food producers in order to prevent the adulteration of food products.[7] According to a historical account offered by Edith and Ochubiojo,

> *These were groups of tradesmen of particular specialties whose purpose was to provide control and general supervision over the honesty and integrity of their members and the quality of their products. In France the most interesting and complete economic document of the Middle Ages available on the subject is the* Livre des Métiers, *which, in the thirteenth century, outlined a code of the comparative practices of the trade guilds of Paris.*[8]

Although private forms of food regulation at the local or national level is not new, it has been limited in scope and authority. Food has always been at the core of states' sovereign autonomy, and governments have played a central role in regulating food mainly to maintain national security and legitimacy. Indeed, the initiative role in the control over food safety and quality has been primarily played by local authorities, city governments, or state agencies, especially in the early stages of urbanization and farming development. Therefore, private actors have usually played a complementary role in the regulation of food at the local and national levels.

Nevertheless, over the past two decades, TPR of food safety has emerged as a new governance sphere and proliferated at the national but more so on the transnational levels.[9] Key drivers of food safety TPR include the growing impact of international trade and the WTO, regulatory failures of national and international institutions, global sourcing by multinational food companies, frequent food safety scandals and resulting public distrust toward governments, and increasing competition among giant brand-name industries. The various TPR schemes have increasingly gained important influence on the market structure as well as risk regulation along the global food supply chain.[10] Such TPR schemes at different levels, with different designs, have given rise not only to concerns but also theoretical and practical questions. The questions have important public health, international trade, and development implications.[11]

This chapter endeavors to explore the structure and processes of TPR in global food safety governance. It offers a brief account of the development and proliferation of TPR as a renewed approach to food safety regulation beyond the

---

7. Radomir Lasztity, Marta Petro-Turza and Tamas Foldesi, *History of Food Quality Standards*, in Encyclopedia of Life Support Systems (EOLSS) (Radomir Lasztity ed.), Developed under the Auspices of the UNESCO (2004).

8. Ihegwuagu Nnemeka Edith and Emeje Martins Ochubiojo, *Food Quality Control: History, Present and Future*, in Scientific, Health and Social Aspects of the Food Industry (Benjamin Valdez ed.) 421, at 423 (2012).

9. *See* Spencer Henson, *The Role of Public and Private Standards in Regulating International Food Markets* (IATRC Summer Symposium: Food Regulation and Trade: Institutional Framework, Concepts of Analysis and Empirical Evidence, Bonn, Germany, May 28-30, 2006 Working Paper).

10. *See* Fabrizio Cafaggi, *Product safety, private standard-setting and information networks*, in *The Regulatory Function of European Private Law* (F. Cafaggi and H. Muir Watt, eds.), at 215 (2009).

11. As regulatory institutions in global administrative space do not operate in isolation, it is crucial to understand how global food safety governance is configured and reconfigured through complex interactions. Such public-private regulatory interactions will be analyzed in detail below.

national and international levels. This chapter asserts that the development is in part due to the ineffectiveness and inefficiency of TPR's public counterparts in coping with this global problem. In the context of the emergence of TPR in food safety, this chapter further looks into the merits and problems of TPR schemes and their possible roles in global food safety governance.

THE DEVELOPMENT OF TPR IN GLOBAL FOOD SAFETY GOVERNANCE

To be sure, some basic forms of TPRs arise through contracts in ordinary international commercial transactions.[12] Importers and retailers in their everyday practice specify in procurement contracts the basics between the seller and buyer, including, for instance, the name, label, quantity, and price of a given product to be purchased as well as shipping and insurance terms. Such contractual practice has developed into more and more complex and sophisticated affairs because of changing market power, business structure, manufacturing techniques, transportation improvement, consumer demand, economic development, and cooperation history. Nowadays, it is normal for procurement contracts to cover public law requirements of a product's destined jurisdictions and detailed specifications of safety and quality requirements such as tolerated amounts of filth and contamination.[13]

From an efficiency perspective, importers and retailers have an incentive to invest resources in designing complex and self-benefiting procurement contracts (and sometimes methods of verification). In the past, however, the enforcement of such contractual requirements was usually underdeveloped and subject to various types and levels of monitoring rigorousness, from no monitoring, consignment, or verification, to business plant inspection. The bottom line is, this basic contractual structure embedded in the ordinary commercial practice in international food trade is able to serve as a means to either export public law requirements beyond the jurisdiction of importing countries or for multinational food companies to develop and expand more demanding private rules and standards to regulate upstream suppliers.

The most prominent example among various private governance institutions is GlobalGAP, a European-based retailer-led organization that sets Good Agricultural Practice (GAP) standards for agricultural products beyond the traditional state-centric, command-and-control sanitary regulations. First established as EurepGAP by retailers of the Euro-Retailer Produce Working Group in 1997, GlobalGAP rebranded itself to reflect its global reach in 2007.[14] GlobalGAP is now the largest private standard-setter of food safety and sustainability in the world, with over 112,600 producers in more than 100 countries under its certification. The GlobalGAP network has thus become deeply rooted in the setup of international food trade.[15]

---

12. *See* Private Food Law: Governing Food Chains through Contract Law, Self-Regulation, Private Standards, Audits and Certification Schemes (Bernd van der Meulen ed.), 77-78 (2011).
13. *Id.*
14. GlobalGAP History, at http://www.globalgap.org/uk_en/who-we-are/history/.
15. *Id.*

Global Sourcing, Local Consumption

Many factors have stimulated and facilitated the development and institutionalization of GlobalGAP and TPR schemes as such.[16] First, the significantly industrialized and globalized patterns of food production and consumption have rendered food safety problems a regulatory issue beyond the scope of a traditional state-centric model. The rise of TPR (e.g., GlobalGAP) can be, to a degree, attributed to the globalization of food production and consumption. Global sourcing has also raised efficiency drivers of multinational food companies to allow these companies to effectively manage and control their upstream suppliers in different countries via TPR—standardization, certification, and third-party auditing methods.

Globalized Food Supply Chains and Outbreaks

Second, globalized food supply chains have intensified the seriousness, scale, frequency, and impact of food scares in the past decades. The gravity of these new situations have, in turn, not only heightened consumer concerns over food safety and distrust of government oversight, but also forced governments as well as the industry to respond to public criticisms by reviewing and reforming existing regulatory mechanisms.

Reputational Risks to the Food Industry

Dreadful food safety scares have raised considerable waves of consumer concerns about food safety. The concerns have further amplified associated reputational costs (risks) on branded food suppliers, especially those in the retail sector.[17] This heightened consumer awareness prompted by the numerous food safety crises spread across the globe together with other food safety issues (of varied risks) such as the use of pesticides or GMOs and hormones and chemicals as animal food additives. This transformed consumer mindset has incentivized TPRs like BRC and EurepGAP to fill the regulatory gaps left by their public counterparts.[18] Many other TPR schemes have also rapidly emerged and are being driven by transnational food corporations' economic incentives in avoiding related reputational damage.[19] Apart from reputational costs concerns,

16. *See e.g.* Tetty Havinga, *Transitions in Food Governance in Europe from National Towards EU and Global Regulation and from Public Towards Hybrid and Private Forms of Governance*, Nijmegen Sociology of Law Working Papers Series 2012/02, 6-9, available at http://ssrn.com/abstract=2189478.

17. *See* Havinga, *supra* note 16, at 6-7.

18. Donal Casey, *Three Puzzles of Private Governance: GlobalGAP and the Regulation of Food Safety and Quality*, UCD Working Papers in Law, Criminology & Socio-Legal Studies Research Paper No. 22/2009, at 10 (2009).

private actors along food supply chains have increasingly engaged in games of product differentiation and marketing strategies.[20]

Public Distrust of Governmental Regulatory Reforms and a Shift of Responsibility

Moreover, there has been a public perception of the ineffectiveness of state-centric regulatory systems, triggering various regulatory governmental reforms with experimental regulatory designs. Recognizing the limits of the traditional command-and-control approach in the global food supply chain era, a few reform projects adopted by governments (especially in Europe) have gradually shifted the responsibility for ensuring food safety from national governments to the food industry. The U.K. Food Safety Act 1990[21] appears to be the benchmark legislation of "devolved responsibility."[22] Food suppliers bear full responsibility for food safety incidents unless they can prove that they have taken all reasonable precautions and have exercised all due diligence to prevent such incidents.[23] This shift of responsibility from the public to the private domain has prompted the development of TPR.

The development and institutionalization of TPR well illustrates "governance beyond the state,"[24] which describes new modes of governance designed and implemented by various nonstate actors. Traditionally, food safety control measures have been regarded as a regulatory space exclusively filled by states, with indirect participation by (the regulated) food companies. Now, these companies are creating TPR schemes that engage in setting and implementing private rules and standards along with overseeing inspection and certification. In the area of food safety governance, TPRs have become critical transnational intermediaries linking and mostly harmonizing domestic regulatory regimes, international

---

19. Tetty Havinga, *Private Regulation of Food Safety by Supermarkets*, 28(4) Law & Pol'y 515, 522-24 (2006) (describing the British Retail Consortium food safety standard adopted by British and Dutch retailers, the International Food Standard adopted by German and French supermarkets, and the newly developed Global Food Safety Initiative, which serves to bring international retailers into agreement on globally accepted food safety standards). *See generally* Neil Gunningham et al., *Social License and Environmental Protection: Why Businesses Go Beyond Compliance*, 29 Law & Soc. Inquiry 307 (2004) (describing the phenomenon of corporations conforming to societal pressures as "social license").
20. More specifically, driven by a variety of consumer expectations and preferences, TPRSs now reach beyond safety and food quality, to environmental protection, fair trade, labor rights, or animal welfare, etc. (a broad scope—not limited to food safety). *See generally* Douglas A. Kysar, *Preferences for Processes: The Process/Product Distinction and the Regulation of Consumer Choice*, 118 Harvard Law Review 525 (2004).
21. *See generally* Colin Scott, "Continuity and Change in British Food Law" 56 Modern Law Review 785, 792-94 (1990).
22. *See* Casey, *supra* note 18, at 7-10.
23. Section 21 of the Act imposes "due diligence" requirements on the ultimate suppliers of branded food products—namely the retailers. Section 21, UK Food Safety Act 1990.
24. *See generally* Colin Scott, *Spontaneous Accountability*, *in* Public Accountability: Designs, Dilemmas and Experiences 174 (Cambridge University Press, 2006).

standards, and rules over production processes in different jurisdictions. The expansion and institutionalization of TPRs have blurred the traditional divide between rule-makers and rule-takers, especially in an extraterritorial context. Increasingly, these TPR schemes erode the centrality enjoyed by state authorities with their individual and combined effects. Many observers argue that this rapid evolvement reflects a paradigm shift from a state-centric approach to a broader focus of governance,[25] as private governance is increasingly gaining influence in the global market. As these private actors gain influence, their authority and sanctions powers derive less and less from national governments.

STRENGTHS OF TPR OF FOOD SAFETY

TPRs in general have several comparative advantages in providing food safety governance as opposed to their public counterparts. Notably, TPR excels in technical expertise, effective supply chain control (especially in a cross-border setting), regulatory adaptivity, and efficiency. However, before jumping into the elaboration of such comparative advantages, the limits of the generalizing discussion below need to be underlined. In fact, there exist hundreds of TPR schemes in food safety governance,[26] varying in their nature, scope, regulatory methodologies (e.g., the use of monitoring and verification such as third-party audits or record-keeping requirements), ideology, decision-making process, governance structure, participating stakeholders, and the operating market configuration. Therefore, the comparative advantages discussed below do not readily apply to all existing TPR schemes as a matter of course.

Technical Expertise

Food corporations usually possess scientific and technical expertise as food technology is both complex and rapidly evolving. For example, genetically modified crops,[27] meat products from cloned animals,[28] and other kinds of novel foods are normally beyond the traditional expertise of the government

25. *See e.g.* Casey, *supra* note 18, at 10 (2009); *more generally, see* Gerry Stoker, *Governance as Theory: Five Propositions*, 50 Journal of Social Sciences 17 (1998).
26. Spencer Henson and John Humphrey, *The Impacts of Private Food Safety Standards on the Food Chain and on Public Standard-Setting Processes*, report prepared for the Joint FAO/WHO Food Standards Programme, Codex Alimentarius commission 32nd Session (June 29-July 4, 2009).
27. *See generally* Guy Cook, Genetically Modified Language: The Discourse of Arguments for GM Crops and Food 78 (2005); Paul St. Amand, *Risks Associated with Genetically Engineered Crops*, *in* Genetically Modified Crops: Their Development, Uses, and Risks 351, 351-52 (G.H. Liang & D.Z. Skinner eds.), (2004); Wolfgang van den Daele, *Risk Prevention and the Political Control of Genetic Engineering: Lessons from a Participatory Technology Assessment on Transgenic Herbicide-resistant Crops*, *in* Genetically Modified Foods: Debating Biotechnology 242, 243-46 (Michael Ruse & David Castle eds.) (2002).
28. FDA has issued a statement declaring "that food from cattle, swine and goat clones is as safe to eat as the food we eat every day." Andrew Martin, *F.D.A. Says Cloned Animals Safe to Eat*, *N.Y. Times*, Jan. 15, 2008, *at a1*, http://tinyurl.com/b7vbfh.

agencies that are charged with administering health and safety regulations. But TPR schemes usually possess their own laboratories, teams of expert scientists, and specialized knowledge and information in food technology, which state institutions may lack. TPRs also have greater "working knowledge of technical terms and basic engineering considerations" than their public counterparts do.[29] Government agencies "frequently rely on private product-safety standards to give their own regulations more credibility" based upon such technical expertise rationale.[30] It is therefore common practice for government agencies to incorporate, modify, or make reference to privately developed standards or approaches to food safety regulation in forming public regulatory measures.[31]

Effective and Efficient Cross-Border Supply Chain Control

In addition, actors participating in the TPR are experienced in creating effective and efficient management, implementation, monitoring, certification, and sanctions in a cross-border context so as to structure how upstream suppliers manufacture food products.[32] TPR is reinforced by the enormous market power possessed by its participants (notably giant multinational retailers through procurement decisions), which implement and enforce TPR rules and standards in more effective and efficient ways than those imposed by the retailers' public sector counterparts.[33]

Although TPR rules and standards are not required by national or international law, they have arguably become de facto mandatory.[34] Theoretically, TPR rules and standards are not legally binding to suppliers—they are required to merely comply with the public laws and regulations of their respective

29. Ross E. Cheit, Setting Safety Standards: Regulation in the Public and Private Sectors 196-202 (1990).

30. Timothy D. Lytton, *Competitive Third-Party Regulation: How Private Certification Can Overcome Constraints that Frustrate Government Regulation*, 15(2) Theoretical Inquiries in Law 539 (2014), at 564.

31. *See e.g.* Peter L. Strauss, *Private Law Organizations and Public Law*, Wm. & Mary Bill Rts. J. (Forthcoming), available at http://papers.ssrn.com/sol3/papers.cfm?abstract_id=2194210; Cheit, *supra* note 30, at 225.

32. David Antony Detomasi, *The Multinational Corporation and Global Governance: Modelling Global Public Policy Networks*, 71(3) J. Bus. Ethics 321, 325 (2007).

33. TPR provides efficient and effective safety control systems along the supply chain for prompt reaction to food safety outbreaks. For instance, when a food safety outbreak triggers a product recall, either voluntarily made by the responsible producer or demanded by the government, TPR actors are generally supported by a well-designed control network along the supply chain that may provide effective risk management. Fabrizio Cafaggi, *Transnational Governance by Contract: Private Regulation and Contractual Networks in Food Safety*, *in* Private Standards and Global Governance: Economic, Legal and Political Perspectives (Axel Marx, Miet Maertwns, Johan Swinnen, and Jan Wouters eds.), 195, at 197 (2012).

34. *See* Gretchen H. Stanton, *Food Safety-related Private Standards: The WTO Perspective*, *in* Private Standards and Global Governance: Economic, Legal and Political Perspectives 235, at 246 (2012).

jurisdictions or export destinations. But as a practical matter, some TPR rules and standards have become the industry norm along global food supply chains. It in effect presents as "binding requirements" when economic power asymmetry exists between the procurer and supplier. Because retail markets in most industrialized countries are highly integrated—especially in Europe and the United States where their respective giant retailers are expected to dominate the global market in the future[35]—the economic power enjoyed by giant multinational corporations leaves little room for developing countries' suppliers. They usually must either comply with TPR or exit the market.[36]

Regulatory Adaptability

TPR also enjoys greater regulatory adaptability. By contrast, public treaties, laws, regulations, and standards often "tend to stay fixed in their original form."[37] TPR is not limited by the regulatory rigidity faced by public institutions.[38] TPR schemes usually have more working knowledge about the rapidly changing food technology, supply chain structure, and market practice. TPR schemes also have to compete with each other for business clients that ask for inspection and certification as well as for consumers that look for credible brand names. TPR schemes have to regularly review and revise (when necessary) their rules, standards, and implementation methodologies in response to changing circumstances and experience.[39] All in all, TPR regulatory adaptability is an important strength, enabling itself to sufficiently adjust to the rapidly evolving science and technology in the area of food safety.

Cooperative and Proactive Nature

TPR is generally of a cooperative rather than coercive or adversarial nature. As explained by Timothy Lytton in discussing third-party certification as a form and an important part of TPR, "[r]egulation by private certification facilitates

35. Doris Fuchs and Agni Kalfagianni, *The Democratic Legitimacy of Private Authority in the Food Chain*, *in* The Challenges of Global Business Authority: Democratic Renewal, Stalemate, or Decay? 65 (Tony Porter and Karsten Ronit eds.), at 71-73.

36. *See* Miet Maertens & Johan Swinnen, *Private Standards, Global Food Supply Chains and the Implications for Developing Countries*, *in* Private Standards and Global Governance: Economic, Legal and Political Perspectives 153, at 153-54 (2012).

37. Cheit, *supra* note 30, at 204.

38. Lytton, *supra* note 31, at 566 (citing Peter H. Schuck, Why Government Fails So Often and How It Might Do (A Little), Chapter 6 (2014)).

39. *See e.g.* Foris Fuchsm Agni Kalfagianni, and Tetty Havinga, *Actors in Private Food Governance: The Legitimacy of Retail Standards and Multistakeholder Initiatives with Civil Society Participation*, *in* Agriculture and Human Values 1, 10-12 (2009); Yesim Yilmaz, *Private Regulation: A Real Alternative for Regulatory Reform*, Cato Policy Analysis No. 303 (April 20, 1998), available at http://www.cato.org/publications/policy-analysis/private-regulation-real-alternative-regulatory-reform; Cheit, *supra* note 30, at 203-18.

cooperation by regulated entities. Within ongoing private agency-client relationships, manufacturers are eager to satisfy the demand of certifiers upon whom they rely for the marketability of the manufacturers' products."[40] Unlike government-led public regulations, inspections, and enforcements (such as issuing warning letters, revoking facility registration, and criminal charges), "[c]ompanies come to see private agencies as marketing partners rather than police."[41]

Such a close agency-client relationship has bearings on the "proactive" feature of TPRs. Consider a case between exporting country X and importing country Y. From a public regulatory perspective, before food products actually arrive at country Y's border inspection posts, country X is not required by law (international, or the relevant national laws) to inform country Y of the exported products' sanitary condition unless there is a multilateral or bilateral international treaty specifying such a requirement. Therefore, government agencies are usually reactive in food safety problems—intervening only when an outbreak triggers public concern.[42] In contrast, food safety TPR possesses important advantages in regulating cross-border food safety practices. TPR actors, whether they be located in country Y or not, with considerable market power are able to *proactively* regulate the behavior of suppliers in country X or other jurisdictions, across the entire supply chain.

WEAKNESSES OF TPR OF FOOD SAFETY

Food safety TPR, like every other regulatory model, is not perfect. Despite the effectiveness, efficiency, expertise, adaptability, and proactivity that TPR may have to offer, it in some circumstances suffers from numerous institutional design problems. Such myriad drawbacks may limit the ability of TPR to develop into a sound global governance system in food safety.

Agency-Client Relationship, Checklist Governance, and Second-Level Information Asymmetry

In an ideal market, both suppliers and buyers are fully informed in making decisions to exchange homogeneous products. In reality, however, goods are not homogeneous and buyers are not always well informed. This is particularly true for the food sector characterized by varying risks, complex processing, biotechnology, and global sourcing. While some qualities are easily ascertained (e.g., shelf life, appearance, smell, or freshness), whether a food product is "safe" (or possessing other "credence qualities"— such as GMO, cage-free, organic, and pesticide-free) cannot be easily known by consumers or procurers at the time of

40. Lytton, *supra* note 31, at 34.
41. *Id.*
42. *See* Lawrence Busch, *Quasi-States? The Unexpected Rise of Private Food Law*, *in* Private Food Law: Governing Food Chains through Contract Law, Self-Regulation, Private Standards, Audits and Certification Schemes 51 (Bernd van der Meulen ed.) (2011), *supra* note 12.

purchase.[43] Such information asymmetry is one of the root causes for the market failure in contemporary food safety regulation, as the public authority does not have the resources to test comprehensively and food suppliers are incentivized to go easy with safety to minimize production costs. Therefore, large retailers require growers and producers to be inspected and verified to meet safety standards[44]; and consumers look for reliable certification labels as quality signals while grocery shopping. The purpose behind the requirement is to avoid legal liability, reduce reputational risks, and also to fuel product sales. Third-party audits and certifications, whose mechanism is to reduce such information asymmetry,[45] serve as both a safety and quality control as well as a marketing tool.

But in some circumstances, third-party audits and certifications as a way of food safety TPR may fail to alleviate such information asymmetry and even create a problematic state that this chapter calls "second-level information asymmetry." Second-level information asymmetry exists where the buyer uses the auditor to cure the first-level information asymmetry between the buyer and the supplier, but in the end, fails to even know if the auditor lives up to his responsibilities. There are two main causes for such a problem: Conflict of interest in the agency-client relationship as well as checklist governance due to system immaturity or ossification.

First, as some commentators have pointed out, there is a problematic "agency-client relationship" between the third-party auditing firm and the company to be inspected.[46] As the company to be audited selects the auditor and pays the certification fees,[47] and numerous auditors are competing for clients, it is rational to expect that auditing firms would cut corners instead of living up to the highest safety standards.[48] As clearly pointed out by Timothy Lytton in his studies of the Kosher certification industry in the United States, "[m]arket competition among certifiers sometimes leads them to lower their standards in order to reduce the cost of their services and ease the demands that they place on their clients."[49] There is, in effect, a race to the bottom due to a conflict of interests and market competition. Critics point out that private organizations, created primarily to realize economic ends, cannot also be neutral contributors to or provider of public goods.[50]

---

43. Friederike Albersmeier *et al.*, *The Reliability of Third-Party Certification in the Food Chain: From Checklists to Risk-Oriented Auditing*, 20 Food Control 927, 927-35 (2000).

44. *See e.g.* Havinga, *supra* note 19.

45. Albersmeier *et al.*, *supra* note 44, at 927-35.

46. *Id.*

47. Elizabeth Weisse, *Food Safety Auditors Are Often Paid by the Firms They Audit*, USA TODAY (Oct. 4, 2010), at http://usatoday30.usatoday.com/money/industries/food/2010-10-01-foodaudits01_ST_N.htm.

48. Albersmeier *et al.*, *supra* note 44.

49. Lytton, *supra* note 31, at 3.

50. Although it remains unclear whether vehicles of a TPRS are ever able to serve as a neutral contributor to or purveyors of public goods, when the narrow interests of shareholders and the welfare of the general public come into conflict, it is reasonable to assume that private corporations would be strongly inclined to privilege the former at the expense of the latter. *See* Detomasi, *supra* note 33, at 326 (referring to Donna J. Wood, *Corporate Social Performance Revisited*, 16 Acad. Mgmt. Rev. 691, 693 (1991)).

Second, given the rapid but relatively young expansion of the third-party audit and certification business, there might not be enough well-trained and experienced expert auditors in the industry.[51] There are some large and established, albeit still young, TPR institutions such as BRC (created in 1992 but the BRC Technical Standard was not adopted until 1998),[52] GlobalGAP (initiated in 1997 as EUREP and modified to GlobalGAP in 2007),[53] and the Global Food Safety Initiative (GFSI) (developed in 2000 by a group of multinational retailers and manufacturers).[54] These TPR institutions may be able to regularly host training sessions, conferences or workshops, and even publish books and journals to support training and continuous professional education in the technically demanding regulatory business. However, because most TPR schemes are still in their early stages of development and in rapidly expanding markets, others may necessarily need to hire freelancing private firms (which may not be subject to stringent and rigorous training and review) to subcontract the auditing and certification tasks. This may result in the crucial problem of "checklist governance"[55] where unprofessional auditors simply stick to a formalistic checklist[56] without paying attention to material deficiencies or special risk circumstances of the industry/sector. Ill-trained or freelancing auditors may fail to have a complete familiarity with raw materials, food additives, and processing techniques utilized in a variety of food plants. The problem of checklist governance might also appear when the day-to-day practice of an auditing third party becomes ossified. Both system immaturity and ossification may lead to regulatory failure that surfaces as "second-level information asymmetry."[57]

---

51. Ching-Fu Lin, *Challenges to Third-Party Food Safety Audit and Certification*, Bill of Health (Dec 5, 2013).

52. For more discussion, *see* Filippo Arfini and Maria Cecilia Mancini, *British Retail Consortium (BRC) Standard: A New Challenge for Firms Involved in the Food Chain. Analysis of Economic and Managerial Aspects*, Institute of Agrifood Economics, Department of Economic and Quantitative Studies, Parma University, available at http://www.origin-food.org/pdf/db/it_brc.pdf.

53. For more discussion, *see* Antoine Bernard de Raymond and Laure Bonnaud, *Beyond the Public-Private Divide: GlobalGAP as a Regulation Repository for Farmers*, 21(2) Int'l J. of Sociology of Agriculture and Food, 227 (2014).

54. For more discussion, *see* Elizabeth Ann Driscoll, *Neoliberal Policy in Action: A Critical Analysis of the Global Food Safety Initiative and Food Safety Governance* (University of Toronto, 2012), available at https://tspace.library.utoronto.ca/bitstream/1807/33646/1/Driscoll_Elizabeth_A_201211_MA.pdf.

55. Albersmeier *et al.*, *supra* note 44, at 927-35.

56. For business efficiency reasons—the corporate motive to minimize costs and to maximize profits is likely to generate resistance to governance mechanisms that add burdens to companies' coffers. *Id.*

57. The recent Jensen Farms salmonella outbreak (2011) in the United States adequately illustrates the problem of "second-level information asymmetry." For more discussion, *see* Dan Flynn, Jensen Brothers Take Responsibility But Blame PrimusLabs, Food Safety News (Oct. 21, 2013), at http://www.foodsafetynews.com/2013/10/jensen-brothers-sue-primus-over-third-party-audit-they-say-was-faulty/#.Up4svGRDtyg; Gabriella Agostinelli, Jensen Farms Sues Food Safety Auditor (Nov. 22, 2013), at http://www.foodlawfirm.com/2013/11/jensen-farms-sues-food-safety-auditor/.

Legitimacy and Accountability Concerns

Another dimension of concern is that TPR traditionally lacks the legitimacy to exercise governmental enforcement authority in the public realm.[58] The considerable structural power TPR schemes exercise along the global supply chain justifies subjecting them to an examination of legitimacy. It is especially true when such TPR schemes are not merely commercial practices but a governance space of rules and standards that have a wide and profound impact on the allocation of authority and resources in the political framework of the global food system.[59]

Indeed, TPR actors do not have that sort of "democratic legitimacy" derived from an electoral mandate or the representativeness enjoyed by their public counterparts.[60] In most cases, the corporate institution, unlike the governmental machinery, is not designed and geared toward the formulation of public policies and rules that take into account the problems of the society as a whole.[61] As opposed to their government counterparts, TPR auditors and certifiers are not democratically legitimate and accountable in the way that citizens in a democracy are used to. Some commentators have also opined that TPRs usually do not have sufficient and equal stakeholder participation "by all groups affected by these [TPR] standards."[62] Furthermore, TPR schemes are also criticized for the lack of an adequate level of transparency, where the decision-making or standard-setting institutions do not fully disclose the processes and material information to the regulated under the TPR framework.[63] As a result, some argue that "most private food standards primarily reflect the interest of retailers" and that "our analysis of the democratic legitimacy of private retail food governance, then, is not an optimistic one."[64] Accordingly, the food safety TPR may suffer from a legitimacy deficit, primarily due to private actors' narrow pursuit of corporate profits rather than public goods[65] along with their lack of an electoral mandate or democratic representativeness as enjoyed by public institutions.[66]

---

58. *See* Detomasi, *supra* note 33, at 325.
59. Fuchs and Kalfagianni, *supra* note 36, at 73.
60. For more discussions of the legitimacy of private ordering, *see e.g.* Lisa Sharma, Stephen Teret, and Kelly Brownell, *The Food Industry and Self-Regulation: Standards to Promote Success and to Avoid Public Health Failures*, 100 American Journal of Public Health 240 (2010).
61. Donna J. Wood, *Corporate Social Performance Revisited*, 16(4) The Acad. Mgmt. Rev. 691, 691-718 (1991).
62. Foris Fuchsm Agni Kalfagianni, and Tetty Havinga, *Actors in Private Food Governance: The Legitimacy of Retail Standards and Multistakeholder Initiatives with Civil Society Participation*, *in* Agriculture and Human Values 1, 10-12 (2009).
63. *Id.*
64. *Id.*
65. Donna J. Wood, *Corporate Social Performance Revisited*, 16(4) The Acad. Mgmt. Rev. 691, 691-718 (1991).
66. *See* Detomasi, *supra* note 33, at 325.

Nevertheless, the lack of representative election or democratic accountability does not necessarily render TPR illegitimate as a governance model. Whether a regulatory institution, public or private, has an adequate level of legitimacy varies considerably among different institutional designs and models and is subject to a case-by-case evaluation based on sufficient empirical evidence. The readily assumed regulatory legitimacy from public mandates in public governance institutions is limited to functioning democracies. Yet even democratic public governments are often troubled with regulatory capture (especially by the food industry) and do not fare much better than TPRs in terms of participation, transparency, and accountability.[67] As pointed out by Timothy Lytton, in the realm of government regulation, for instance, "notice-and-comment rulemaking by [U.S.] government agencies can be heavily influenced behind the scenes by industry stakeholders."[68] The whole rule-making process, as analyzed by other commentators, in effect reflects the imbalance between industry and consumer groups in a way in favor of the industry interest."[69] Therefore, even public governance institutions may best be accountable to only a fraction of the people affected by their regulatory activities. When viewed on the global scale, it is unclear whether public governance institutions can be equally legitimate and accountable beyond mere theoretical assumption.[70]

Finally, the interactive and polycentric nature of global food safety governance may call for a new understanding of legitimacy. Institutions in interactive food safety governance may become less important as regulators execute their mandates and exercise their authority, but more important as *platforms* for interinstitutional interactions.[71] While the idea of global food safety governance confers no supremacy to either public or private institutions, their legitimacy derives from the extent to which their institutional design embeds social norms in the global marketplace.[72] As the new models of global food safety governance

---

67. Lytton, *supra* note 31, at 36.
68. *Id.*
69. Wendy Wagner, Katherine Barnes, and Lisa Peters, *Rulemaking in the Shade: An Empirical Study of EPA's Air Toxic Regulations*, 63 Administrative Law Review 119 (2011).
70. In many aspects, such as supply chain expertise, participation, accountability, responsiveness, adaptability, and transparency, TPR institutions often perform better than the multi-level, bureaucratic public institutions. Therefore, TPR often secures a certain level of legitimacy by filling the gap where their public counterparts cannot fill. This is particularly true when there is currently no agreed understanding of legitimacy that is readily appropriate and sufficient for evaluating TPR, or "governance beyond the state" (most common criteria proposed by scholars, for example, have included concepts of input legitimacy, output legitimacy, and deliberative democracy). For more on the comparison between GlobalGAP and Codex, *see* Ching-Fu Lin, *Public-Private Interactions in Global Food Safety Governance*, 69(2) Food and Drug Law Journal 143 (2014).
71. *See* discussion below.
72. That is, norms "that derive authority directly from interested audiences, including those they seek to regulate, not from sovereign states." Steven Bernstein and Benjamin Cashore, *Can Non-State Global governance be Legitimate? An Analytical Framework*, 1 Governance and Regulation 347, 348 (2007).

increasingly "derive their legitimacy from the voluntary and conditional participation of individuals who can revoke their consent at any time,"[73] the understanding of legitimacy of global governance may need to be reconceptualized.

Distributional Impact on Small Farmers and Developing Countries

Last but not least, TPR has raised concerns over its distributional impact, especially on small farmers in developing countries. Demanding TPR schemes have emerged from developed countries and have considerably affected developing countries' exports, resulting in developing countries criticizing TPRs as new sanitary and phytosanitary (SPS) market barriers that undermine their trade (export opportunities) and rights to development.[74] Some argue that "[w]hile some positive effects on food safety and some quality can be recognized in Western countries, domestic food quality in developing countries has not benefited accordingly as these positive effects are often restricted to food products for export markets…"[75] Moreover, there has been in developing countries "a trend toward the marginalization of small farmers and retailers, and subsequently an increase in economic inequality due to the stringent retail standards."[76]

For many developing countries, TPR in the food sector has caused problems for it conflicts with public rules and standards in many ways. First, due to the technical and administrative complexity associated with TPR, which provides detailed and prescriptive requirements for every stage of the production and processing chain—"down to the level of each individual producer or processing plant"[77]—the costs of compliance may be high and, consequently, some developing country suppliers (especially small farmers) may be unable to afford them.[78] Second, developing countries often complain that many TPR standards seem to exceed relevant government requirements.[79] Third, the proliferation of TPR has translated into possibly overlapping and even conflicting public and private requirements.[80] There are numerous TPR standards that differ widely across countries, regions, and products, and they remain fragmented and far from being universally

---

73. James Rosenau, Distant Proximities: Dynamics Beyond Globalization 308 (Princeton University Press, Princeton 2003).

74. Maertens & Swinnen, *supra* note 37 at 153-54.

75. Fuchs and Kalfagianni, *supra* note 36.

76. Fuchs and Kalfagianni, *supra* note 36.

77. Stanton, *supra* note 35, at 242.

78. *See* Grace Chia-Hui Lee, *Private Food Standards and Their Impacts on Developing Countries*, European Commission DG Trade Unit G2 (2006), available at http://trade.ec.europa.eu/doclib/docs/2006/november/tradoc_127969.pdf.

79. Some exceeding standards include maximal levels of pesticide residues of 50% or even 33% of public standards established by the importing countries or much more stringent and complicated animal disease control steps. Stanton, *supra* note 35, at 246.

80. "At the time, the United Nations Conference on Trade and Development (UNCTAD) estimated the number of private schemes at 400 and increasing." *Id.*, at 239.

applicable. As TPR in general does not possess the institutional competence to ensure comprehensive monitoring and enforcement, its actual practice may unnecessarily increase the costs of compliance, especially for small producers.

CONCLUSION

TPR functions with the economic power of multinational corporations in highly integrated markets, which renders TPR de facto mandatory. Increasingly, exporting-country suppliers cannot distinguish between private and public requirements as they are generally unaware of the nature, source, or process of "law" in an importing country as well as the so-called public-private divide. Such potential regulatory competition is further amplified by the fact that the public-private interface is increasingly blurred especially in the cross-border setting. Public and TPR mechanisms, as a result, overlap as potentially competing regulatory spheres at the transnational level. Such overlap and blurred boundaries have raised concerns in interactions between public and TPR sites of governance in the global food regulatory space, which is of crucial importance but underanalyzed.[81] To fill this gap, this chapter examines two important dimensions of interaction between public and TPR regulatory mechanisms in global food safety governance and discusses possible future development in global food safety governance through constructive interactions.

Interactions Between TPR and National Authorities

Given the ineffectiveness of public institutions in ensuring the safety of food products along global supply chains, TPR stands out for its adaptability, effectiveness, and transnational nature. As opposed to the time and costs reflected in the negotiation, adoption, and revision of multilateral international treaties, TPR has an important strength in that it possesses flexibility, which enables its regulatory mechanisms to adapt to the rapidly evolving area of food technology.[82] It is against such a background that an increasing number of states have encouraged the involvement of private actors in food safety regulatory

81. The term "global food regulatory space" here falls in the concept of "global administrative space," as developed by Benedict Kingsbury, Nico Krisch & Richard B. Stewart, *The Emergence of Global Administrative Law*, 68 L. & Contemp. Probs. 15 (2005). There are some studies on TPR food standards, such as those on GlobalGAP's impacts or the WTO SPS Agreement's provide possible responses to TPR standards. However, such studies tend to observe TPR institutions (*e.g.*, GlobalGAP) as well as public institutions (*e.g.*, WTO) in isolation, and overlook the theoretical and practical implications of studying how interactions take place between the two. *See e.g.* Stanton, *supra* note 35; Denise Prevost, *Private Sector Food-Safety Standards and the SPS Agreement: Challenges and Possibilities*, 33 South African Yearbook of Int'l Law 1 (2008).

82. For example, genetically modified crops, cloned meat products, and many other novel foods are usually beyond the traditional expertise of the government agencies, even those with health and safety mandates. Transnational food corporations and industry associations, on the other hand, usually possess their own laboratories, teams of expert scientists, and specialized knowledge and information in food technology, which public institutions may lack.

processes (albeit in a hierarchical way). Examples include the food safety regulatory mechanisms incorporating or partnering with TPR, such as those in the U.S. Food Safety Modernization Act (FSMA)[83] as well as that of the EU General Food Law.[84] This complementary public-private interaction between TPR and national authorities has again reinforced the development and institutionalization of TPR instruments.[85]

However, for many developing countries, TPR has caused unnecessary problems for it competes with public rules and standards in many ways. As previously discussed, such problems include the technical and administrative complexity and high costs of compliance, more stringent and complicated control steps and standards than public ones, and possibly overlapping and even conflicting public and private requirements. As a result, because such demanding TPR requirements have posed considerable developmental and distributional issues, some developing countries, which rely on food export, have criticized TPR for undermining their market access and development, while others have responded by adapting to TPR.

On the one hand, a few developing countries have come to recognize the dominance of TPR in global food supply chains as a market reality and have responded by cooperating with TPR schemes. For example, Kenya, Chile, India, Malaysia, and Mexico[86] have developed government-led GAPs and certification services at very low or no cost to assist small suppliers. Furthermore, they have benchmarked such public certification services against private standards so as to be recognized as equivalent to GlobalGAP or other TPR schemes.

On the other hand, some developing countries have brought the issue of TPR standards up to the international level. Developing countries have used such international platforms to challenge TPR and have requested that food regulation stay within public control or, at the very least, that TPR be disciplined by principles set by public institutions in a hierarchical manner. For example, in 2005, a delegate from Saint Vincent and the Grenadines brought a complaint to the WTO SPS Committee about GlobalGAP requirements that restricted the country's ability to export bananas to the EU.[87] By raising its concern about

---

83. *E.g.* FSMA Foreign Supplier Verification Programs. Food Safety Modernization Act (FSMA) Title III, Sec. 301. The FSMA also requires that importers verify that their foreign suppliers have implemented proper preventive mechanisms. *Id.*

84. Regulation (EC) No. 178/2002 of 29 Jan. 2002, 2002 O.J. (L 31/1), available at http://eur-lex.europa. eu/LexUriServ/LexUriServ.do?uri=OJ:L:2002:031:0001:0024:EN:PDF [hereinafter EU General Food Law].

85. This however makes some (consumer) NGOs raise concerns that private modes of governance will undermine efforts to induce governments to regulate, or detract from the credibility and reliability of having a "public" inspection or process to ensure compliance.

86. Olga van der Valk & Joop can der Roest, *National Benchmarking Against GlobalGAP: Case Studies of Good Agricultural Practices in Kenya, Malaysia, Mexico and Chile* (LEI Report, 2009), available at http://edepot.wur.nl/11453.

87. WTO SPS Committee, *Private Sector Standards Discussed as SPS Committee Adopts Two Reports*, WTO: News Items (June 29-30, 2005), at http://www.wto.org/english/news_e/news05_e/sps_june05_e.htm.

GlobalGAP and other TPR schemes in general at the WTO, Saint Vincent and the Grenadines triggered heated debate at the international level with countries divided on the issue.[88] Similar contested discussions also took place in other intergovernmental organizations, including inter alia the Codex, the World Organization for Animal Health (OIE), and the Organisation for Economic Co-operation and Development (OECD).

Interactions Between TPR and International Organizations

The position of Saint Vincent and the Grenadines was supported by Jamaica, Peru, Ecuador, and Argentina. Because the issue of TPR has since then been raised frequently as a specific trade concern at the SPS committee, the SPS committee jointly established a workshop on private and commercial standards with the UN Conference on Trade and Development (UNCTAD) Trade & Sustainable Development, International Trade Division in June 2007. The joint workshop reviewed different TPR schemes, conducted case studies on many developing countries, and discussed the implications of TPR standards in terms of market access, development issues, and WTO rules.[89]

In the SPS committee meeting on April 2-3, 2008, WTO members and the OIE focused largely on the interface between TPR and public standards. First, the director-general of OIE warned that standards set by GlobalGAP or other TPR actors may undermine the science-based and democratically adopted standards of public international organizations and may therefore pose difficulties for developing countries.[90] Moreover, OIE's submission to the SPS committee emphasized that TPR has not been examined against the SPS agreement, and "there is reason to believe that many private standards are not consistent with SPS obligations."[91] As the OIE was seriously concerned about the potential trade-restricting and -distorting effects of private standards, it proposed that the SPS committee act as a forum for countries to identify and discuss specific problems with TPR.[92]

In the SPS committee, Uruguay and Egypt headed a group comprised of developing countries that argued that TPR rules and standards are "arbitrary and can be difficult for developing countries to meet." WTO members are obliged under the SPS agreement to discipline their nongovernmental bodies to respect SPS rules.[93] They in particular based the arguments upon Article 13 of the SPS

---

88. Stanton, *supra* note 35, at 235.
89. Joint UNCTAD/WTO Informal Information Session on Private Standards (June 25, 2007), at http://www.wto.org/english/tratop_e/sps_e/private_standards_june07_e/private_standards_june2007_e.doc.
90. The public international organizations which play the role as international standard-setters in the WTO SPS contexts are the Codex, the OIE, and the International Plant Protection Convention (IPPC).
91. Considerations Relevant to Private Standards in the Field of Animal Health, Food Safety and Animal Welfare (Submission by the OIE, February 25, 2008, SPS Committee G/SPS/GEN/822).
92. *Id.*
93. WTO, *Sanitary and Phytosanitary Measures: Members Set to Agree on Regionalization, Improved SPS Transparency*, WTO: News Items (April 2-3, 2008), at http://www.wto.org/english/news_e/news08_e/sps_apr08_e.htm.

agreement, which reads "Members shall take such reasonable measures as may be available to them to ensure that non-governmental entities within their territories, …comply with the relevant provisions of this Agreement."[94] Others rebutted this argument claiming that the "nongovernmental entities" in Article 13 cannot be interpreted to include GlobalGAP and TPR schemes alike, as at the time of the Uruguay Round negotiations, there was no such phenomenon of TPR and the drafters of the SPS agreement were not aware of this dramatic development and the institutionalization of TPR. It is therefore argued that governments have no obligation to ensure the compliance of private actors and their contractual commercial terms and conditions with the SPS agreement.[95]

On May 29, 2009, the 76th General Session of the OIE International Committee adopted a resolution on the implications of TPR standards in international trade of animals and animal products.[96] The resolution asked the director general of the OIE "to work with relevant public and private international organizations with the objective that concerns of Members are taken into consideration and that TPR standards, where used, are consistent with and *do not conflict with those of the OIE* [emphasis added]."[97] In the same year, the 32nd session of the Codex meeting discussed a report on TPR food standards prepared by Food and Agriculture Organization (FAO) consultants, which systematically compared different types of TPR standards against the presumption that they pose challenges to the Codex.[98] The 2009 Codex Report suggested that Codex include TPR standard bodies as observers to Codex meetings, and reform Codex procedures to better respond to the various challenges of TPR standards. However, many developing countries rejected the 2009 Codex Report findings and insisted that Codex stand as "the sole international reference organization for food safety."[99] During the 33rd session of the 2010 Codex meeting, another report was considered that highlighted many developing countries' concerns, one being that TPR standards may undermine the Codex's authority and other official requirements, that third-party certification is technically and financially difficult, that the TPR standard-setting process generally lacks a sound scientific

94. Article 13, Agreement on the Application of Sanitary and Phytosanitary Measures, Annex 1A, The Legal Texts: The Results of the Uruguay Round of Multilateral Trade Negotiations 59 (2000), 1867 U.N.T.S. 493 (1994) [hereinafter SPS Agreement]. "In addition, Members shall not take measures which have the effect of, directly or indirectly, requiring or encouraging such regional or non-governmental entities, or local governmental bodies, to act in a manner inconsistent with the provisions of this Agreement" *Id.*

95. WTO, *supra* note 95.

96. *Implications of Private Standards in International Trade of Animals and Animal Products*, OIE Resolution No. XXXII adopted by the 76th General Session of OIE International Committee, at http://www.oie.int/fileadmin/Home/eng/Internationa_Standard_Setting/docs/pdf/en_resolution_20N_C2_B0_20XXXII.pdf.

97. *Id.*, para. 3.

98. *See* Henson and Humphrey, *supra* note 27.

99. *See* Stanton, *supra* note 35, at 248.

basis and transparency, and that TPR standards have negative impacts on market access and development.[100] The 2010 report nevertheless favored the stance that Codex standards, given they are scientifically backed and democratically adopted, should be the "benchmarks" for their TPR counterparts.

In the same year, the OIE organized a meeting with TPR standard-setting bodies, including GlobalGAP and GFSI. OIE meeting attendees agreed that "the basis for private standards on sanitary safety are the existing international standards of OIE and Codex Alimentarius, as well as national and regional legislation," and that future collaboration between OIE, GlobalGAP, and GFSI should be encouraged.[101] Subsequently, in May 2010 the 78th General Session of the OIE International Committee passed another resolution that requested the OIE "maintain and strengthen appropriate links and dialog with relevant global private standard setting bodies and global private industry organizations with the aim to allow *compatibility* of private standards with OIE standards."[102] Most importantly, toward this objective, in another meeting, the OIE followed up with selected TPR bodies with a global reach and explored the possibility of signing official agreements with GlobalGAP and GFSI. The OIE concluded an official agreement with GFSI, as it is more of a benchmarking institution than a standard-setting institution.[103] However, because GlobalGAP is a standard-setting body and potentially competes with the OIE, the OIE decided not to enter into any official agreement with GlobalGAP at that time, but rather to merely continue to share information under an exchange-of-letters mechanism.[104]

Governance Through Constructive Interactions

The interactions between public and TPR institutions to date have been limited. The perspective of the public institutions discussed here is rather one-way, hierarchical, and state centric.[105] Reactions at the national level have been polarized based upon nation states' respective interests and concerns—from trade and development to health protection. In the WTO context, although Article 13 of the

---

100. *Consideration of the Impact of Private Standards*, Agenda Item 13 (CX/CAC 10/33/13) for the Joint FAO/WHO Food Standards Programme, Codex Alimentarius commission 33rd Session (July 5-9, 2010).

101. OIE, *Implications of Private Standards in International Trade of Animals and Animal Products*, at http://www.oie.int/international-standard-setting/implications-of-private-standards/.

102. OIE, *Roles of Public and Private Standards in Animal Health and Animal Welfare*, OIE Resolution No. 26 adopted by the 78th General Session of OIE International Committee, at http://www.oie.int/fileadmin/Home/eng/Internationa_Standard_Setting/docs/pdf/A_RESO_2010_PS.pdf.

103. OIE, *Report of the Meeting of the OIE Animal Production Food Safety Working Group*, (November 15-17, 2011), http://www.oie.int/fileadmin/Home/eng/Food_Safety/docs/pdf/A__Final_Working_Group_Report_November_2011.pdf.

104. It is further noted that "there are fundamentally different approaches to animal welfare standards of the OIE and GlobalGAP (the latter's standards being largely based on European Union legislation, which is overly prescriptive compared with the approach accepted by OIE Members)." *Id.*

105. As we can clearly see that the Codex 2010 Report underlines the "benchmark" superiority of Codex, as well as the OIE Resolutions, which treat GlobalGAP from the perspective of competition rather than of cooperation.

SPS agreement leaves some room (albeit little) for the argument of disciplining TPR standards against international trade rules, it remains unclear if and if so, to what extent should WTO members be required to "take such reasonable measures as may be available" to manage the dynamics of TPR in the food safety arena. The relationship between TPR represented by GlobalGAP and international public institutions as discussed above is not a traditional one, and the public institution's use of legal interpretation as a means to claim supremacy over or eliminate overlaps with TPR seems to be of very limited utility. In the Codex and OIE contexts, attempts to create and define boundaries between public and TPR authorities based on the concepts of competence, expertise, or legal mandate may not work in such unconventional relationships. Nevertheless, while the empirical evidence on how TPR affects developing countries is limited and the debates remain,[106] and although the public health benefits of TPR has not been well considered, it is too early for relevant public institutions to "eliminate potential competitors" before seeking interinstitutional conversation and collaboration.

As a matter of fact, GlobalGAP and other TPR schemes will continue to expand the range and variety of rules and standards to cover not only food safety standards, but also standards for sustainability, labor rights, and animal welfare.[107] As it seems unrealistic to try to contain and isolate food safety TPR or ignore its influence in the global market, more constructive interactions are desirable. TPR interactions with its public counterparts are constructive to global food safety governance. As no meta-framework exists or appears to be emerging that dictates, coordinates, or pioneers the development of such diverse norms, the public as well as TPR institutions together constitute an evolving governance complex being formed and transformed, configured and reconfigured by different actors at different levels with different experimental approaches.[108] Interactions between TPR and public institutions in the process can serve as a vehicle for constructivist learning, mimesis, experimentation, and innovation, projecting normative effects on participating governance sites.[109] Dynamic interactions between TPR and public institutions can also help build scientific networks, activity coordination, information exchange, joint projects, and expert collaboration, which are mutually reinforcing in strengthening global food safety governance.

---

106. Thomas Herzfeld, Larissa S. Drescher, and Carola Grebitus, *Cross-National Adoption of Private Food Quality Standards*, 36(3) Food Policy 401 (2011).

107. WTO, *supra* note 95.

108. Ching-Fu Lin, *Reassessing the Codex Alimentarius Commission: An Analysis of the Ractopamine Hydrochloride Dispute*, Paper to the Future of Transnational Law: The EU, USA, China, and the BRICS (2012 WISH Conference), Shenzhen, China (2012).

109. The fact that GlobalGAP spurs regulatory reforms in many countries (both developing and developed), that the OIE takes a step toward legal innovation by signing an official cooperation agreement with a NGO (IGO's signing official agreements with NGOs in the food safety arena is a legal innovation through inter-institutional relations, although in some other areas there have been a couple of "public-private partnership" projects between the WHO and pharmaceutical companies), and that the Codex reassess its decision-making process against the arguably more effective private standard-setters all well demonstrate constructive interactions.

Index

Note: Page numbers followed by *b* indicate box.

D

H

I

FSC
www.fsc.org
MIX
Papier | Fördert gute Waldnutzung
FSC® C083411

Kolibri 360 GmbH
Ferdinand-Jühlke-Straße 7
99095 Erfurt, Deutschland
produktsicherheit@kolibri360.de